DEVELOPING the PACIFIC NORTHWEST

Developing the Pacific Northwest

The Life and Work of Asahel Curtis

William H. Wilson

Washington State University
Pullman, Washington

Washington State University Press
PO Box 645910
Pullman, Washington 99164-5910
Phone: 800-354-7360
Fax: 509-335-8568
Email: wsupress@wsu.edu
Website: wsupress.wsu.edu

First printing 2015

Library of Congress Cataloging-in-Publication Data

Wilson, William H. (William Henry), 1935-
Developing the Pacific Northwest : the life and work of Asahel Curtis / William H. Wilson.
pages cm
Includes bibliographical references and index.
ISBN 978-0-87422-331-6 (alk. paper)
1. Curtis, Asahel, 1874-1941. 2. Photographers--Northwest, Pacific--Biography. 3. Mountaineers--Northwest, Pacific--Biography. 4. Conservationists--Washington (State)--Biography. 5. Roads--Environmental aspects--Washington (State)--History. 6. Mount Rainier National Park (Wash.)--History. 7. Olympic National Park (Wash.)--History. 8. Washington (State)--Environmental conditions--History. 9. Civic leaders--Washington (State)--Biography. 10. Ranchers--Washington (State)--Yakima River Valley--Biography. I. Title. II. Title: Life and work of Asahel Curtis.
TR140.C818W37 2015
770.92--dc23
[B]

2015011459

On the cover: From Asahel Curtis's whimsical series of photographs of young students from Seattle's Mary Ann Wells School of Dance, pictured in Mount Rainier National Park in September 1931. *Washington State Historical Society, Tacoma, 1943.42.58589.*

Publication of this book was supported by a generous grant from

Furthermore: a program of the J. M. Kaplan Fund

Table of Contents

Acknowledgments

YEARS AGO A. Theodore "Ted" Brown mused about the way historians have thanked their wives or husbands when acknowledging their help in completing a book. He suggested that someone should write an article on the subject. Brown was then the director of the History of Kansas City Project and I was one of the project graduate students probing the past of Kansas City, Missouri. Among the vast number of matters I'd never thought about was how historians thanked their spouses, but Ted's remark spurred me to pay more attention to acknowledgment pages, where I found wives, husbands, and sometimes families the last to be thanked, bringing up the rear as putative stragglers or afterthoughts, no matter how much they were praised. No one ever wrote Ted's proposed article, as far as I'm aware, but a recent recollection of it prompts me to break the mold by thanking my wife, Katharine L. Wilson, first.

Beyond her always steadfast support and encouragement, Kitty made four direct contributions to this book. First, she sympathized with my failed efforts to find out the mark or brand of the autos Curtis used as conveyances and props for the "nymphs," the dance school pupils he photographed at the Sunrise area of Mount Rainier National Park. She reasoned that, since Curtis made the Rainier trip in September 1931, the 1932 models could have been delivered by then. Sure enough, the cars proved to be 1932 Nash automobiles, which never occurred to this self-imagined "car guy." Second, she urged me to discover whether a Curtis collaborator, Eula Lee Merrill, had papers that would shed more light on his efforts to preserve or replant trees, shrubs, and flowers along Washington roadsides. The Merrill Papers at the University of Washington did reveal much about Curtis's concerns. Third, she patiently listened to me read the entire manuscript twice, each time offering wise suggestions for improvement. Fourth, she shared her knowledge of photography and photo developing, essential to a study of Asahel Curtis.

Every member of the staff of Special Collections, University of Washington Libraries, was courteous and helpful despite my repeated requests to examine this or that box of the Asahel Curtis Papers, the principal source of information on Curtis's activities. Gary Lundell deserves a mention by name because of his mastery of collateral material in Special Collections and his unfailing interest in advancing any work dependent on sources in Special Collections. At the Research Center of the Washington State Historical Society, Tacoma, Joy Werlink displayed her remarkable knowledge, not only of the Curtis photographs housed there, but also of the archival material available to researchers. Her knowledge is exceeded only by her patience. Several collections in the Washington State Archives, Olympia, where Lupita Lopez kindly unearthed sources and bore the brunt of my requests, were important to this study. At the Puget Sound Regional Branch of the state archives in Bellevue, Midori Okazaki found the Kitsap County property records detailing the disposition of the Curtis farm in the Port Orchard area. At the University of Alaska, Rosemarie Speranza led the way to the Archie Satterfield Collection, a trove compiled by one of Curtis's biographers. Kenneth Howe and especially Patty McNamee of Seattle's Pacific Alaska Region, National Archives and Records Administration, helped identify pertinent material in the records of the Mount Rainier National Park administration and other sources. Ruth Steele of the Center for Pacific Northwest Studies, Western Washington University, Bellingham, provided minutes, reports, and letters from the records of the Washington State Good Roads Association, as well as those of the North End Improvement Council, contained within the records of the good roads organization.

Some events surrounding the sale of Curtis's "ranch," then west of Grandview, Washington, may always remain unexplained. To the extent that I was able to penetrate the mysteries of the sale, I am grateful for information provided by Diane Guthrie of the Sunnyside Valley Irrigation District, Sunnyside; Brigid Clift of the Central Regional Branch of the Washington State Archives, Ellensburg; Jennifer Richter of the Fidelity Title Company, Yakima; and the staff of the Auditor's Office, Yakima County Courthouse. All of them took an exceptional interest in getting to the bottom of the sale, even though I was unable to reach all the way despite their help. An interview with Richard Herriman, at his farm adjacent to the former Curtis property, enriched my understanding of farming in the area even though I did not use any of the interview material in this book.

The staffs of the University of Washington's Suzzallo-Allen, Health Sciences, and Engineering Libraries, and of Seattle's Central Library probably regarded their assistance to that nondescript patron to be routine, but to me it was cherished and indispensable.

Small portions of Chapter Three first appeared in "'Names Joined Together as Our Hearts Are': The Friendship of Samuel Hill and Reginald H. Thomson," *Pacific Northwest Quarterly* (Fall 2003), while the themes of Chapters Nine and Ten originally were developed in "Asahel Curtis and the Fight over the Olympic National Park," for the same journal in its summer 2008 issue. I am grateful to Kim McKaig, the editor of the *Pacific Northwest Quarterly*, for permission to use that material here. Gratitude is also extended to Christina Orange DuBois, the managing editor of *Columbia: The Magazine of Northwest History*, who granted permission to use, in Chapter Three and in altered form, the essential argument of "The War League: Asahel Curtis's Plan for Peace," published in *Columbia's* winter 2009-10 issue. All three articles are cited in the list of sources.

At the Washington State University Press I am indebted to Robert A. Clark, editor-in-chief, and to Beth DeWeese, manuscript editor. Their initial and continuing faith in the manuscript was most welcome. Later, Beth's informed, thoughtful, and careful editing made for a much better final version. I am also in debt to Edward Sala, director; Kerry Darnall, copyeditor; Nancy Grunewald, designer; and Caryn Lawton, marketing manager.

Three anonymous readers for the Press offered comments and suggestions that significantly improved the manuscript. I am especially grateful to the reader who revealed a vital source, the online historical archive of the *Seattle Times*, in the course of making many wise observations in a friendly manner.

Kitty's and my younger daughter, Margaret, helped, as she always does, with medical questions based on her nursing experience.

At the end of the day, none of these fine people is responsible for the errors or inadequacies of this book. They are my responsibility alone.

Abbreviations in the Text

AAA	American Automobile Association
AASHO	American Association of State Highway Officials
Argus	*Seattle Argus* (weekly)
BPR	Bureau of Public Roads
ECC	Emergency Conservation Committee
LHA	Lincoln Highway Association
NEH	National Endowment for the Humanities
NEIC	North End Improvement Council (Kitsap County, Washington)
P-I	*Seattle Post-Intelligencer*
PWA	Public Works Administration
RNPAB	Rainier National Park Advisory Board
RNPC	Rainier National Park Company
Times	*Seattle Daily Times, Seattle Sunday Times*
WSCPRB	Washington State Council for the Protection of Roadside Beauty
WSGRA	Washington State Good Roads Association
WSHS	Washington State Historical Society
WSPC	Washington State Planning Council
YTA	Yellowstone Trail Association

Curtis's family poked fun at his careless dress but this portrait of a mature Curtis shows a man very careful about his attire. *Washington State Historical Society, Tacoma, 1943.42.2013.0.3.*

INTRODUCTION

The Contradictions of Asahel Curtis

ASAHEL CURTIS did not drive a car.[1] Yet for thirty years he fought to bring first-class highways to Washington State through the "good roads" movement, a determined effort to persuade state and federal governments to finance, build, and constantly improve a road network. He believed that tourists and Washington's citizens could best savor the state's visual delights while passing through them or reaching them by car. Only broad, smooth highways could carry them in comfort and safety. Only good roads could reach into every area of the state, stimulating commerce and industry while increasing the knowledge and appreciation realized through first-hand experience. Curtis (1874-1941) lived during the era of widespread rail travel, so his aversion to driving presented few barriers to movement even when he could not ride with others. At home, his wife Florence was the family chauffeur.

Though his home for most of his adult life was in Seattle, Curtis was an avid mountain climber and outdoorsman who thrived on roughing it in the boondocks. Once, during a month-long expedition on the Olympic Peninsula, he forsook a feather bed to sleep among the animals in his hosts' barn. Yet he campaigned for tourist roads in Mount Rainier National Park while heading a citizen consulting group, the Rainier National Park Advisory Board. He wanted some roads built on ridge tops because he knew that most visitors to the park arrived in cars or buses and never wandered far from their vehicles. The park's spectacular scenery should be available to the vast majority of park goers, not just to people like him, the small minority of physically fit, active hikers and horseback riders. Opposition from other members of the minority resulted in fewer roads than he wished, and those that were constructed sometimes failed to follow his desired routes.

Although rightly regarded as a premier Northwest photographer who took, or assigned his staff to take, tens of thousands of photographs, Curtis was not obsessed with his profession. Many of his photographs of

Mount Rainier, of other mountains, and of varied scenes in Seattle and elsewhere are beautifully composed, yet he refused to consider himself an artist or to discuss photography as an art. He took precious time and energy from his business to pursue his interests in Rainier, good roads, and his nine and one-tenth acre "ranch" in eastern Washington's Yakima Valley. What the Curtis family called a ranch was really a small orchard in most of its length and breadth. Curtis, whose humor was sly, could have chosen the informal name as a facetious reference its modest acreage. His other concerns included the future of the Olympic Peninsula and planning for the Columbia River basin. Somehow he worked in dozens of magazine and newspaper articles, talks, and speeches. The main function of the Curtis studio, then, was to support him, his other interests, and his family, whether or not he consciously willed that purpose.

Indeed Asahel Curtis had a family. He married in 1902 and was married to the same woman at his death almost thirty-nine years later. Four children were born to them. There was no hint of infidelity by Curtis in the marriage, and no suggestion of any scandalous sexual behavior before he married, even during the two years he spent in the unbridled atmosphere of the Alaska-Yukon gold rush. In the phrase of the day, Curtis was a "family man." On the other hand, the family enjoyed little emotional closeness. Of the four children, only his older daughter shared her father's struggle to keep his business afloat during the Great Depression of the 1930s while working as his secretary. Soon after his return from Alaska to Seattle, Curtis became alienated from his (subsequently) more famous photographer brother, Edward, the creator of the magnificent *North American Indian* pictorial series. The two rarely spoke. Their self-imposed distance included their wives and children, who were unacquainted although they lived in the same city for many years.

Short of stature, about five feet, six inches tall, Curtis was as unimpressive personally as he was physically. A shy man, he was bereft of small talk, at least on first acquaintance. He possessed a lively sense of humor but it was usually well concealed beneath a layer of apparent diffidence. At the same time he was a compelling speaker, as when, in 1928, he held the "rapt attention" of 350 banqueters at a national convention during a forty minute illustrated talk on the industries and scenic attractions of Washington.[2] His personality encouraged comradeship more than friendship, though he demonstrated forceful leadership in dangerous situations on the slopes of the Olympic Mountains and, later, in trying circumstances on the heights of Rainier. His shyness proved no barrier to blunt attacks on others when significant issues provoked his wrath. His opponents in the fights over renaming Mount Rainier to Mount Tacoma and establishing a gigantic Olympic National Park could testify to his uninhibited ire.

Curtis was a fervent conservationist and a militant preserver of beautiful natural landscapes and natural resources. Yet he also enthusiastically supported the exploitation of the natural resources of Washington State and the Pacific Northwest. His contemporary critics did not understand how he could embrace both resource preservation and resource development, and so assumed that he was a hypocrite or a toady of the exploiters. Later detractors piled on, condemning his developmentalism and damning his conservationism and preservationism as inadequate or misdirected. They expressed no sympathy when Curtis had to confront conflicts between his principles, when circumstances forced him to face inescapable contradictions and choose one principle over another. That circumstance scarcely made him unique among human beings. In the case of preservation versus development, Curtis was clear about which principle he chose and why, a clarity that upset his critics all the more.

A devoted Republican, Curtis was a political and social conservative who was firmly entrepreneurial in outlook. He had no patience with able bodied men who refused to work in return for public support during the Great Depression. Yet he belied the stereotype of the benighted reactionary when he firmly defended state and federal public works as legitimate responses to widespread unemployment. He advocated federal intervention in the rural economy through land reclamation and its related activities including the federal government's valuation of reclaimed land, federal dam construction, and the organization of cooperative irrigation districts to which farmers were obliged to belong. He enthusiastically supported interstate compacts to develop flood control, pollution abatement, and land use restrictions along the Columbia and other rivers. He premised his leadership of the good roads movement on persuading governments to tax and spend steadily increasing sums for highways.

These contradictions rested comfortably in the Curtis psyche, at least most of the time. Together they formed an individual more complex, more fascinating than either his opponents or defenders would concede.

Illuminating or resolving Curtis's contradictions is a daunting task because his surviving correspondence bunches in the years from his middle forties to his middle sixties. During these years his views changed little, the one notable exception being his altered outlook on the National Park Service and the U.S. Forest Service. The Forest Service rose in his estimation from the 1920s onward because he believed that its policy of mixed use of federal land, blending preservation, recreation, and development, was best for his state. Beginning in the 1930s he viewed the Park Service less as an agent of preservation than as an expansionist bureaucracy devoted to aggrandizing land and resources at the expense of controlled economic development. As these examples illustrate, usually he saw the issues confronting him through the lens of the best advantage to the state of Washington and, to a lesser extent, the Pacific Northwest.

Statewide unity and harmony, statewide preservation and development, statewide economic and population growth led, he believed, to improvement in the human condition in Washington. On these matters his papers adequately represent his mature views. On others, including the development of his belief system, or family and personal relationships, there is next to nothing. His daughters purged his papers before donating them to the University of Washington (see Author's Note), and few personal references escaped their attention. Reminiscences or anecdotes about Curtis date almost exclusively from his later adult years. His children's reflections on such concerns as his estrangement from his brother Edward, his business ability, and the loss of his Yakima Valley farm, are either conjectural or misinformed.

Beyond these issues, Curtis compartmentalized his life to an extraordinary degree. To be sure, there was some overlap in his widely varied activities. His interest in good roads was intertwined with his interest in opening Mount Rainier National Park to automobile tourists, while his concern for good roads and the improved marketing of farm products identified with his irrigated land near Grandview. Fundamentally, however, only his photography embraced all of his disparate pursuits, as his wildflower pictures connected with his veneration of Mount Rainier.

The chapter organization that follows, while roughly chronological, adheres to the distinctly topical grouping of Curtis's activities. Chapter One reviews his early life and his adventures before, during, and after his return from the Yukon gold rush. His photography is the focus of

Chapter Two, although the chapter considers other matters such as Curtis's boosterism, and his personality and family life as well as they may be recovered from the materials available. Chapters Three, Four, and Five span the good roads movement and his intense involvement in many of its ramifications, including his presidency of the Washington State Good Roads Association. Most impassioned of all was his long and deep association with Mount Rainier, the Mount Rainier National Park, and the Rainier National Park Advisory Board, the subjects of Chapters Six, Seven, and Eight. His positive outlook on the future of the Olympic Peninsula, based on his extensive knowledge of its mountains and valleys, is explored in Chapter Nine. His enthusiasm for the peninsula's economic opportunities explains his single minded opposition to what became the Olympic National Park, to him a vast federal sanctuary that foreclosed the legitimate exploitation of some of the area's abundant resources. Chapter Ten explains and analyzes his stand against the park. The topic of Chapter Eleven, the eastern Washington ranch, overlaps much of Curtis's post-Yukon life, revealing some of his positive personal qualities while highlighting an experience enigmatic in its origins and conclusion. Chapter Twelve looks at Curtis's later years, and offers some judgments about the purpose and meaning of his life.

Loggers worked hard and Curtis respected hard work, but he could not resist a good-humored parody of the woodsmen's habit of having their photos taken while recumbent in the notches of soon-to-be-felled forest giants. In this 1910 photo Curtis wears a world-weary expression, as though he himself had easily notched the tree with the axe. *Washington State Historical Society, Tacoma, 1943.42.18746.*

Chapter One

Growing Up, Joining the Gold Rush, and Returning

Asahel (pronounced A-shul in the family) Curtis first saw the Puget Sound country in 1888, when he was thirteen years old. He, his mother, Ellen Sheriff Curtis, and his older sister Eva, traveled from rural southern Minnesota that spring. They went to join Johnson Asahel Curtis, Ellen's husband. They did not have much money, so they probably met their Northern Pacific train at the most convenient point on the line. Their train was an emigrant train, the type the western railroads designed to take people who intended to farm out West to their destinations. The fare was cheap because the railroads hoped for a return on the produce from their farms.[1]

The trip was no luxury event. The modest fare bought Spartan accommodations, as Eva remembered them. A coal stove at one end of the car provided what heat there was and a place to cook "provisions from our farm," while at the other end there was one toilet and space for the three of them and other emigrants to wrap themselves in their blankets. The car itself probably was a drafty, wooden relic, a castoff from main line service. It was dangerous because of the stove and its potential for spilling live coals in the event of a wreck or even a bad jar. The weather of the later 1880s was severe, so the stove returned little heat for its possibly fatal presence. High elevations, such as the temporary trestles over Stampede Pass in the Cascade Mountains east of Puget Sound, would have brought penetrating cold into the car despite the stove. Yet to seventeen-year-old Eva the arrangements "seemed like luxury," perhaps because she thought less about the dangers and focused more on the contrast with the horseback and buggy rides of rural Minnesota.[2]

Once in Tacoma the threesome entrained for Seattle on the Northern Pacific's branch line, derided in Seattle as the "orphan road" because of its inconvenient schedule and poor service. From Seattle's Elliott Bay they boarded a ferry, one of Seattle's "mosquito fleet" of small passenger and freight vessels, for the journey west across the sound to little Sidney, later Port Orchard. At Sidney, Johnson and another son, Edward, had built a cabin, planted an orchard, and bought or established a brickyard. The family reunited, except for the oldest son, for three days. Then Johnson Curtis died of pneumonia. He was forty-eight.[3]

The Civil War broke Johnson's health. On the surface, his service was ordinary enough. He enlisted at the town of Whitewater, Wisconsin, during the summer of 1862, leaving his wife, first born son Raphael (called Ray), and his farm. He joined Company D of the Twenty-Eighth Wisconsin Volunteer Infantry Regiment as a private, and a private he remained. After mustering in, training, and helping to put down a small draft riot, the regiment moved to Kentucky, and then to Helena, Arkansas, where it arrived on January 5, 1863. It would stay there for much of the rest of the war. Among other actions, it defeated a larger Confederate force at Helena, then won the battle of Little Rock, both in 1863. Near war's end the unit moved to Alabama, and afterward to Brownsville, Texas, where it was mustered out in August 1865.[4]

All this appears unremarkable for wartime until one considers the situation of the Twenty-Eighth. Most of its service occurred in Dixie, where malaria was endemic.[5] Typically in the Civil War, marches were tiring, while temporary camps lacked comforts and sanitation. The supply of flyblown food varied but tended toward the inadequate. Troops slept in bedrolls during most weather, in uniforms rarely changed, with infrequent bathing, and minimal, aseptic medical care. In short, dirt, mud, dust, filth, and stench were common, along with endless cycles of cold and heat. The squalid routine overcame Johnson at Helena, where he was twice listed as "sick," first "since" the twenty-fourth of February, and again in March 1863.[6] No diagnosis or treatment was recorded. The illness, whatever it was, was more than the sniffles but less than a debilitating disease justifying his separation from the service. Though he remained comparatively healthy afterward, his later disability probably dates from the Helena episode.

The Twenty-Eighth's heaviest engagement was at Helena on July 4, when a numerically superior Confederate force fought to conquer the fortified Union position on the Mississippi River. Troops of the Twenty-Eighth fired from rifle pits to protect an artillery battery. Private

Curtis, sick or not, probably was on duty in a muddy rifle pit, where he and his comrades faced two grim options, kill enough Confederates to stop them, or be overrun, bayoneted while trying to reload their muskets, or taken prisoner. They held. The Confederates retreated, their grip on Arkansas loosened. The simultaneous Union capture of Vicksburg, Mississippi, and the subsequent battle of Little Rock ended serious Confederate action in Arkansas.[7]

Private Curtis was mustered out with his unit. Several biographers of Edward, the more famous son, repeat an unfounded claim that Johnson was a chaplain. Chaplains were commissioned in the Union army and attached to each regimental staff. Had Johnson been promoted to a chaplaincy the action would have been noted in his record. A man with his profound religious convictions could have offered spiritual succor to his fellows, but that did not make him a chaplain. The chaplain fiction may be a family effort to burnish his undistinguished military career, but Johnson's service needs no apology. His honorable duty in a conflict that undermined his health needs no added lustre.[8]

Johnson went home to Ellen, Ray, and his farm. Edward was born there in 1868, followed by Eva in 1870. The family moved three years later to Le Sueur County, Minnesota, to property owned by Johnson's father, also named Asahel. The name Asahel was a Biblical name, not common even in an era of naming children after Bible characters, but ordinary enough for families to pass it down as a first or second name for generations. Thus Asahel was born in Le Sueur County on November 5, 1874. Hardscrabble farming continued until 1880, when the family moved to the settlement of Cordova in the same county. There Johnson, his health continuing its downward spiral, opened a grocery store. By later standards, grocery stores of the era were small, unsanitary, almost windowless, and poorly lighted. There is no evidence that Johnson's was any better, but he struggled along as he had with farming until 1887, when vicious winters and drought devastated the Minnesota countryside. The family continued to farm, at least in a small way; recall Eva's statement about subsisting on their farm's "provisions" during her trip west with her mother and Asahel.[9]

Meanwhile Johnson became a traveling preacher for the evangelical United Brethren Church, with Edward often along to do the packing, canoe paddling, portaging, and cooking. Johnson may have been a member of another church before joining the United Brethren but he apparently always had been a deeply religious Christian. As he had with farming and storekeeping, he earned almost no money from preaching. Usually his flock paid him in farm produce for preaching or presiding at baptisms, weddings, and funerals. Johnson's health issues came to a head in 1887, when he sought a more prosperous life on the shores of Puget Sound. After Johnson's death the next year, Edward became the head of the household in fact. He and Asahel continued to improve the property, hire themselves out to other farms and businesses to raise cash, and tried to make life bearable for Ellen and Eva. All that ended for Edward when he suffered a back injury in 1890 or 1891. Later he moved to Seattle to begin his famous career in photography. Asahel continued working and perhaps finishing out his primary education. Years later he told his daughter Betty that a man who recognized his ability offered to send him to high school, "but he couldn't be spared on the farm."[10]

When he moved to Seattle, Edward chose well, for Seattle was a place to be in the 1890s, a western wonder city like Denver, Omaha, Kansas City, or Los Angeles. Scarcely more than a village in 1880, the rapidly growing settlement suffered a devastating fire in June 1889. Calamitous fires were a staple of nineteenth-century cities, and Seattle's was competitive in its destructive consequences. Despite the heroics of firefighters and other volunteers, the fire fed on plank streets, the undersides of wooden wharves, frame buildings, houses, and the combustible contents of warehouses, hotels, and stores. Stately brick buildings went down before it while firefighters, never well directed, battled on with diminishing water pressure. In some seven hours, 116 acres of the retail-commercial core and nearby residential sites were reduced to smoking, flickering rubble. Just one brick building was saved. Relief efforts began before the smoke cleared and rebuilding began soon after.[11]

For all its destruction, the fire was a boon. It fostered better street alignments, and encouraged raising the grades of some streets originally placed too low for effective sewage removal. It mandated an abundant water supply. The rebuilding and public works boom meliorated the grisly economic depression that began in the early nineties. The Great Northern Railroad arrived in 1893, making Seattle its western terminus, a move forcing the Northern Pacific to improve its Seattle service. The railroads helped to keep the city going economically. Then, in July 1897, the steamer *Portland* arrived from Alaska with its "ton of gold" on board and news of the massive gold strikes in the Yukon. Stampeders flocked to Seattle on the way to the gold fields, fueling an overnight surge in outfitting and less savory businesses. The gold rush triggered a boom in Seattle, the major United States port nearest Alaska and the Yukon, forging links with Alaska that held for over a century. Despite fire and depression, the city grew fantastically, from a mere 3,553 in 1880 to a post-fire 42,837 ten years later. Then it almost doubled in population, to 80,871 in 1900.[12]

Once in lively Seattle, Edward throve on photography and never looked back. Tall, six feet or better, handsome, and charming when he wished to be, he soon became a leading, if not the leading, photographer in Seattle. He climbed and photographed Mount Rainier, activities in which Asahel would surpass him. A tireless self-promoter, he later became the official photographer of the 1899 Harriman expedition to Alaska and the creator of the famed twenty-volume photographic epic, *The North American Indian.*[13]

Before all that happened, Edward moved Ellen, Eva, and Asahel to Seattle. Asahel went to work for his brother, learning photoengraving and photography. Precisely when the family moved is not clear, but Edward, married in 1892, is first listed in the Polk city directory for 1893. In the 1895-96 directory, Asahel, Ellen, and Eva are grouped at Edward's quarters, 1102 Ninth Avenue, on a high ridge north and east of the retail-commercial core. The next year Edward's address changed to 413 Eighth Avenue, closer to his studio. Asahel was listed as living at that address, along with Ellen and Eva, then and during the time he spent in Alaska. His occupation was listed as engraver, and, in 1899, as photographer.[14]

Asahel's separation from Edward began when he journeyed to Alaska and Canada's British Columbia and Yukon in September 1897 to photograph the gold rush. Photography advanced rapidly in the years before 1897 to become a routine, if complex exercise. By then an effective system for capturing a photographic image of a person or any other subject, was the "dry plate." The dry plate process involved coating a glass plate, eight inches by ten inches was among the standard sizes, with light-sensitive chemicals. The plate was protected from light and breakage until it was inserted in a camera. After the photographer opened the camera's shutter and exposed the glass, the resulting image was again protected from light and breakage until it could be developed into a negative in a mild acid bath. Next, the photographer placed a piece of light-sensitive paper on the negative, and briefly shined a light through the paper, transferring the negative on the glass into a positive image that mimicked the tonal values of the subject. The photographer placed the paper in a developing bath to bring out the image's proper shading, and finally into a solution designed to "fix" or stop the development. Then the paper was hung to dry. The entire process had to be protected from extraneous light, thus the need for a darkroom with only a dim red light to illuminate the work.[15]

If all this seems clumsy and complicated, it was a breeze compared to the tyrannies of the preceding "wet plate" process. The wet plate system was next to impossible on the rough, remote trails of British Columbia and the Yukon because it required the tedious mixing of chemicals and applying them to a glass the night before the photo shoot. The plate had to be developed quickly or its image would fade, requiring a portable but necessarily bulky darkroom near at hand.

Another advantage of the dry plate method was that both the coated glass plates and coated papers could be mass produced. Individual photographers usually developed their own prints, but prints could be mass produced, too. However done, a carefully crafted dry plate print was a remarkably clear, well defined monochrome of delicate shading.

Hand-held cameras existed, but professionals like Asahel disdained them for their fixed-focus lenses, their inherent instability, the relatively poor quality of early celluloid roll film, and the long exposure times that blurred the film if subjects moved. During Asahel's career the developing process simplified, exposure times shortened, most glass plates gave way to better roll film, enlargers became common, hand-held cameras significantly improved, and color photography emerged. Yet the fundamentals remained unchanged from the days of his Alaska-Yukon adventure.

Preparations for that adventure began soon after the *Portland* docked in 1897 with its news of Yukon riches. Confusion, mystery, and controversy surround Asahel's trip to the North. As nearly as the outline of his trip may be recovered, it is this: he left for Skagway, Alaska, a portal to the Canadian gold fields, in September 1897, photographing the gold rushers and doing odd jobs until sometime in 1898. He then went to Dawson City and the adjacent gold fields, where, with one or two partners, he staked a claim on Sulphur Creek southeast of Dawson.

Asahel left the Klondike in 1899, probably during the late summer.[16] The confusion arises from Edward's claim that he, Edward, traveled to Alaska and the Yukon in 1897. The evidence is overwhelmingly against him. No record of him exists in the Yukon Archives Genealogy Research Database, whereas Asahel is listed. A September 18, 1897, article in the weekly *Seattle Argus* (*Argus*) made no statement about Edward's travel to the gold fields but did announce that he would "go into the Alaska view business on the most gigantic scale ever attempted" during the following spring, when he would "start a party for the fields of gold in charge of Mr. A. Curtis."[17]

Some observations about this announcement are in order. It places Asahel's trip in the spring of 1898 but Asahel had in fact left or was about to leave when it appeared. The probable explanation for the discrepancy is that between the time of the article's submission and its publication Edward had decided to send Asahel sooner. The decision would explain why the Curtis family worked frantically to prepare dried and canned food, a tent, and other material for Asahel's early departure.[18] The article's exaggerations about "gigantic scale" and the "party...in charge of Mr. A. Curtis," as well as the assertion that the "party" intended to "take views of sections on which the foot of a human being has never been placed," were pure Edward. Asahel was "in charge" of only himself. The emphasis on a collection of Alaska scenes, "the largest and best that has ever been attempted in this wonderful country" ignored the reality that all of the mining and almost all of the land distance to the gold fields was in Canada. Edward's hyperbole belied the Curtis family's hard work to prepare for Asahel's departure. The food was necessary because Canadian officials announced their determination to let no fortune hunter cross from Alaska to Canada without a year's supply of food and other essentials. The bulk of a year's supply worked out to somewhere between 1,000 and 2,000 pounds. The investment in Asahel was not just physical and emotional, but heavily financial. Hundreds of glass plates, the necessary chemicals, and a good view camera with a tripod had to be assembled and protected from water damage and jostling. All of it had to be moved to wharfside, loaded aboard ship, offloaded at Skagway, and moved to the Canadian line, some twenty miles above Skagway, where the real adventure began.

For all that, it made sense to send Asahel alone. He was a bachelor, young, vigorous, and unencumbered. Preparing him for Alaska and the Klondike taxed the family but their diligence paid off. The family and its descendants made no claim, then or later, of readying Edward for the same trip. Such an effort would have been beyond the family's capability if it went to the limit to organize Asahel's journey. Besides, Edward had his studio to operate, as well as the responsibility for his mother, his wife, their three children, his sister, and two in-laws. A final piece of evidence is an entry in Asahel's Yukon diary dated January 15, 1899: "I write to Ed as plainly as possible regarding the Klondike. Doubt if I will be believed."[19] Had Edward accompanied Asahel, or had Asahel been aware of Edward's article on the gold rush, of which more later, he would not have written that letter.

There was no question about Asahel's going to Skagway, about 1,230 miles north of Seattle by water. The distance from the rambunctious Alaska port to free-wheeling Dawson City at the edge of the gold fields was another 625 miles by land, lake, and river, less than 2,000 miles in all. Contrast this with the "all water route," 2,500 miles from Seattle to the bleak Yukon River port of St. Michael on Alaska's west coast, where the argonauts transferred to shallow-draft river steamers. Then the sternwheelers chuffed 1,723 miles to Dawson, more than 4,000 miles total. Worse, while Asahel was preparing to leave for Skagway, the Yukon was beginning its "freezeup," which meant, for a gold seeker, being marooned in St. Michael or along the wastes of the lower Yukon until "breakup" late the following June. The all-water route was easier going, but the Curtis clan was impatient of delay and Asahel, twenty-two years old, toughened by farm labor and hiking, was inured to the dangers and hardships of the trail. Besides, Edward had made a deal with a leading local newspaper of the day, the *Seattle Post-Intelligencer* (*P-I*), and possibly with the *Seattle Times* (*Times*) as well, to supply gold rush photographs. There would be precious little gold rushing around the Yukon's mouth during the bitter winter of far western Alaska.[20]

So Asahel headed directly north aboard the steamer *Rosalie*, a converted sailing vessel, under the rigorous command of Captain John A. "Dynamite Johnny" O'Brien. "Dynamite Johnny" was aptly named. A vigorous forty-six, a devout Roman Catholic, O'Brien was Irish born and fireplug built, short, heavily muscled, thick bodied, and highly intelligent with a gift of almost total recall. A loyal friend, his reaction to troublemakers such as unruly crew members or anyone who denigrated his religion, was to hit first with his pile driver fists and a stream of profanity that usually excluded blasphemy. Asahel, no troublemaker, was already friendly with O'Brien and with the purser, Charles V. La Farge. More important for Asahel's survival, O'Brien possessed an encyclopedic knowledge of the Inside Passage, its uncharted rocks, its squalls, and its heavy snowstorms. Not least was O'Brien's ability to stay on the bridge during dangerous times, guiding his craft without rest or sleep until smooth sailing returned.[21]

Purser La Farge "proved to be my guardian angel," Asahel later wrote, "and he and Capt. O'Brien showed me the mysteries of navigating the scantily marked channels of the Inland [sic] Passage." At Skagway, shallow waters, no wharf, and the chance of a storm forced shippers to stay well off shore. Crews loaded passengers and freight into light-draft lighters able to strike the beach at high tide, where they stuck for unloading until the next high tide freed them to return to the big ships. Horses were "dumped overboard to swim ashore." La Farge took Asahel aboard the first lighter, but without his camera and probably without the rest of his gear. A storm brewed up while Asahel waited for his equipment. O'Brien, wise in the ways of Lynn Canal weather, retreated with the *Rosalie* from the open roadstead to a safe harbor, from which he returned to Skagway when conditions improved. Once he had camera and supplies in hand, Asahel photographed the wealth seekers and the varied environments they encountered from Skagway to Dawson and the gold fields beyond.[22]

The mysteries of Asahel's Alaska-British Columbia-Yukon adventure involve various incidents in that vast country. Some are known because many of his photographs survive, and his movements may be surmised through them. Others are captured in the reminiscences of another adventurer, in a brief diary that Asahel kept near the end of his stay in the Klondike, and in his own recollections written more than forty years later.

Taking a portion of his known activities first, his venture began as soon as he arrived in Skagway with all of his baggage. Naively, he asked for a hotel room in the swarming town, but was shown a stall barely large enough for a cot. He photographed Skagway but did not stay long. Away from Skagway he took photos of gold seekers, sometimes bearded and haggard, often in deep snow or on rocky ground. They were doing the grim work of moving themselves and their belongings to the gold fields; toiling up the Chilkoot Pass from Dyea, the next port west of Skagway; pausing among rocks for a moment's respite before continuing on; building boats to float down the riverine lakes to Dawson; sailing sled boats along a frozen lake; shooting the treacherous Whitehorse Rapids while surrounded by foam and spray; or camping with other argonauts while waiting for "breakup." The most poignant photos are of horses dead on the White Pass Trail, or abandoned by their owners and soon to be dead in an area of scarce browse. Sharp rocks, boggy ground, and long waits in the line of stampeders, all the time heavily loaded, were too much for most horses. A person does not have to be an animal lover to recoil at the advertising deceptions, ignorance of conditions, and behavior of human beings surrounding the rush to the Klondike.[23]

Beyond the surviving photos practically nothing is known about Asahel's routine, assuming he had a routine. Apparently he went to Alaska with little money—as distinguished from his food and equipment—because one source of income was taking photos of rushers and developing them on postcard stock. The argonauts mailed the cards to relatives back home as proof of their arrival in the storied gold country. Asahel charged $1.50 per postcard, not a bad price when a typical meal at a bare-bones trailside tent restaurant went for $2.50. Similar fare in better surroundings in the states could cost one fourth as much, or less. The number of photographs he took, what he charged for photos that he mailed on behalf of his fellow rushers, or his system for sending or receiving mail, are unknown. He sent photos to Edward but Edward's claim of 150 in the first batch is surely an exaggeration as wild as the other misstatements in his 1897 letter to *The Century* magazine (see below). Asahel served as an unofficial mail carrier for the trail communities, unsurprisingly, given his obligation to return to Skagway periodically and mail his exposed glass plates to Edward.[24]

Glimpses of trail life in Asahel's diary and reminiscences usually come with no specific date and at most a general location. Once he saw a homemade boat swirling and shifting in the dangerous Whitehorse Rapids. He was amazed at the daring and skill of the skipper, a lone, dark figure in the stern. When the boat cleared the rapids and came close to Curtis, he was astonished to see that the "pilot" was a large black dog.

During the fall of 1898 he cautioned a party from Seattle not to attempt a lake crossing on ice as yet thin, urging them to wait until the ice thickened. The men replied that they could not wait; they were in a hurry to arrive in Dawson. Two of the party crashed through the ice and drowned. On another occasion he entered one of the ubiquitous restaurants along the trail, where the proprietor offered him a freshly cut steak. Suspicious, or perhaps merely short of cash, Asahel opted for more modest fare, later to find a dead steer, the source of the meat, by trailside.[25]

The best record of Asahel's abilities and his toughness on the trail comes from the diary and reminiscences of Eugene "Gene" C. Allen, who left Seattle in February 1898 to establish a newspaper in Dawson. He, his brother, and other partners met Curtis, an "old friend from Seattle, . . . already famous as a scenic photographer," a pleasant exaggeration, who was "taking news pictures along the trail." Allen and Asahel decided that Curtis would join the group. Their immediate problem was that they had neither food nor funds to travel farther without a waiver from the Northwest Mounted Police. The issue point for waivers was at the head of Lake Bennett, some 120 miles back down the trail, a daunting trip by foot or dog team. By then the lake and river ice were weakening, adding urgency to Allen's trek. Asahel volunteered to go. He made the round trip in four days, a testimony to his toughness and to the trust the Mounties placed in a well-known man of the trail. The waivers were good only to Big Salmon, a point well short of Dawson, but either the party later sneaked by the Mounties or, what was more likely, the Mounties looked the other way while they journeyed on.[26]

On April 8 they found Fifty Mile River's ice "almost impassible," and "like rotten cloth, and every step we had to watch that we did not plunge to death through an air pocket or a rotten spot." The next day the trio (Allen's friend Joe Dizard was along) with three dogs and their sled, mushed to Miles Canyon, a narrow "hell hole" hemmed in by steep rock walls. They portaged around Miles Canyon. Meanwhile Allen developed a severe case of snow blindness. On the tenth they had to make a final, frightening river crossing involving several trips to move everything from bank to bank. Water rose as high as their knees as they walked across the softening ice below. They barely "made that final trip when the ice went out with a sickening, crushing roar." Allen and Dizard arranged for a raft to take them the rest of the way, and relinquished the dogs and sled to

Asahel to return. Allen went on to establish the first newspaper in Dawson, the *Klondike Nugget.*[27]

Asahel worked on the trail until later that year, then went down to Dawson. Many more photos of Dawson and the gold creeks survive than do his shots of the trail. Miners jam the streets of Dawson, listening to the news of the military victory over the Spanish fleet in Cuba read from a platform, or, on less crowded days, buying food at an open market or selling their leftover goods on the riverside at Dawson after giving up on the Klondike. They crowd the creeks, digging shafts, sawing wood, setting steam points to thaw frozen ground, driving dog teams, and shoveling potential pay dirt into sluice boxes. Sometime in 1898 Asahel staked a claim, number 22059,[28] on Sulphur Creek, 60 above discovery, or the sixtieth claim upstream from the first strike of gold. Southeast of Dawson, Sulphur was not far from Eldorado, one of the fabled creeks of the Klondike, but a long way in riches, at least for Asahel's claim.

By the time Asahel selected his diggings the rich creeks were long since staked. By then, too, Canadian law had reduced the size of a claim from 500 to 250 feet along both sides of a creek, except for the discovery claim. The cross-creek distance still extended from one side of the valley to the other.[29] Curtis and his partner or partners built a cabin, most likely the best part of the nearly valueless claim as it boasted a floor and real glass windows. The windows, a rare luxury, probably came from some of Asahel's less successful or light struck glass negatives.

The claim itself was an apparent money loser despite hard work. At the time that Asahel was active in the diggings miners followed a crude and slow but simple method of reaching bedrock. They selected a likely spot near the creek bank. They stripped the overburden of naturally thawed ground away from an area large enough in diameter to accommodate a miner and his tools. Next they built a fire on top of the permanently frozen ground, hard-as-rock permafrost. When the fire burned out they shoveled the muddy thawed earth, gravel, and miscellaneous debris, piling up a growing mound, their "dump." Then they began over again, driving the shaft deeper with each fire. At bedrock, five feet or so down if they were lucky, they began "drifting," digging side tunnels at right angles to the shaft, hoping to find "colors" or rich deposits in the gravel of ancient stream beds.[30]

All that was winter work, laborious, dangerous, and almost airless once a miner worked his way below the surface. Drifting was intuitive, and frustrating if it yielded little or no gold, frequently the case in the short run. Meanwhile the miners built dams to divert the creeks into their claims when spring or early summer arrived and the flood of snowmelt surged down the creeks. The dams diverted the water into a sluicebox, basically a long wooden trough into which the miners shoveled their dump. The dirt and most gravel washed away while the gold, usually light in weight but of a higher specific gravity, fell behind "riffles" or narrow wooden cross pieces in the sluiceboxes.[31]

Asahel was as gold hungry as the next man, noting no "colors" in mid-November 1898 and "some pay" in a drift tunnel on January 12, 1899. The Curtis crew built a dam and sank two shafts, one of them frequently filling with water from an underground stream. Gathering wood, building fires, scooping away thawed muck during short daylight hours while the thermometer registered in the thirties and forties below zero was too demanding for all day, every day work. Asahel regularly visited other claims, took and developed photographs for "the boys," mailed photos to addresses given him, shot ptarmigan, played cards, and hoped for a job. The potential job was with his friend Eugene Allen, photoengraving for the *Klondike Nugget*. The work would be his salvation because by mid February it was "plain...that things are closing in on me." The job did not materialize, and "I will have to figure very closely to pull through."[32]

The diary ends in February 1899. Asahel may have returned to his claim to work during "cleanup," the spring and summer work of shoveling dirt from the dump into the sluicebox. If he did, the results were dismal. By then it was all over for the Dawson of 1896-1898. Wealthy men were consolidating claims, developing more effective methods of reaching the gold, and, most revolutionary of all, the big strike on the beaches at Nome pulled thousands out of Dawson in the summer of 1899. One of Asahel's final photos is of Dawson's Front Street on May 15, 1899, a short time after an April fire. The snow has melted, the street is dry, the lengthy shadows slanting west to east, and the big clock reading 9:15, all confirm a long, dreamy, bewitching summer evening in an atmosphere so light it seems magical.[33] Unfortunately Asahel was in no condition to rhapsodize over a glorious summer day in the Far North. Only the discovery of a few coins on a riverbank saved him from begging. How he gained the wherewithal to return to Seattle is one of the mysteries of his life. He may have begged and borrowed, or thrown himself on the mercy of friends, or taken a job and saved his money long enough to pay his way out. Edward may have advanced the money, in which case Asahel would not acknowledge his brother's assistance, given their later antagonism. However he arranged his passage, he returned to Seattle. Temporarily, as things developed, he was once more working in Edward's studio.

The discord between Asahel and Edward began after Asahel's return to Seattle, and quickly spiraled into a lifelong estrangement. Much ink has been spilled over this issue, although Edward's primary responsibility for the break is now beyond dispute. Edward fired the opening round in the battle when he wrote a letter to *The Century* magazine on October 14, 1897, although he perhaps did not imagine how it would trap him into justifying a lie. He probably wrote the letter soon after the first batch of photos arrived from Asahel. "I have just returned from a trip over the different trails to the Alaskan gold fields," he wrote in a sentence extremely sparing of the truth. He had 150 views covering every phase of the "mad rush" to reach the gold, including "dead horses literally in piles." Edward was at least honest about the content of Asahel's photographs and his desire to write either an illustrated article or simply a photo essay for the magazine. His concluding sentence is nevertheless stupefying in its departure from reality: "I have myself passed over the trails, climbed the mountains, crossed the rivers, waded through the mud and snow, and endured the cold and hunger."[34] The kindest thing that could be said about this piece of late frontier gasconade is that it worked.

Edward's article, "The Rush to the Klondike Over the Mountain Passes," appeared in *The Century*'s March 1898 issue. The phrase, "pictures from photographs by the author" followed in small capital letters below Edward's name. Whether an editor or Edward inserted the expression is immaterial, because Edward, having claimed personal experience in Alaska and the Yukon, could hardly have repudiated it, even assuming he wished to do so. He had already made a direct, personal claim to creating the seven photographs displayed in his article. Photos aside, the article was a well written, if pessimistic, appraisal of the 1897 rush, based on stories by disappointed refugees from the Far North, backtrackers who failed to survive their first brush with harsh conditions and unrelenting demands. At the close Edward offered some hope that "the signs of victory will be more frequent" in 1898 than in 1897, thanks to improved access. Even so, "the victors will purchase their triumph dearly."[35]

Exactly when or how Asahel became aware of Edward's appropriation of his photographs is unknown. Minor variations aside, the story goes like this: Asahel demanded the ownership of, or, what amounts to the same thing, copyright to the photographs he took in Alaska. His insistence ran counter to the prevailing custom of exclusive company rights to all production from employees of the firm. In this case the "company" was Edward, the sole proprietor, who rejected Asahel's importunate request. The resulting breach never healed. What little contact the two had from then on was almost entirely through third parties. One result of what could be called a negotiable estrangement was Asahel's regaining at least some of his Alaska glass negatives. He scratched off Edward's company signature, replacing it with his own. The families lived separate lives, with the children of Edward and Asahel growing up not knowing their cousins. A telegram to Edward informing him of his brother's death in 1941 went unanswered.[36] The strain on mother Ellen and sister Eva, caught in the middle, can only be imagined.

Except for the realities of the rupture itself, such as cousins raised in ignorance of their relatives, or the unanswered telegram, there are several problems with this story. The first difficulty is that most of it comes from Edward's side of the family. Defenses of Asahel present his side and blame Edward for a lack of fraternal sympathy without disputing existing studio practice.[37] Even granting Edward's prerogatives, a careful biographer of the older brother admits that "[t]he particulars of the falling out are not known."[38] A second problem is the fact of Asahel's residence. In 1900 he was living under Edward's roof at 413 Eighth Avenue in a menagerie comprising himself and nine other relatives.[39] Whatever the specific arrangements regarding Asahel's salary and boarding bill or rent, he was a guest in Edward's home and Edward's employee. It would have been poor form for him to intrude on his brother's sufferance with a demand for individual ownership of company property. Asahel was abundantly capable of anger, as episodes in his later career revealed, but it stretches credulity to believe that he would be so boorish as to begin a family row over a settled commercial practice. Thirdly, and apart from any other issue, Asahel knew the rules of the photography game as well as anyone else. There must be, therefore, some better explanation for Asahel's wrath.

The answer to Asahel's anger is in the words "pictures from photographs by the author" listed below Edward's name in the article. By this phrasing Edward asserted his creation of photos produced by Asahel. Had Edward or an editor written something like "photographs from the E. S. Curtis Studio," or had there been no mention of the source of the photos, Asahel would have had no legitimate complaint. Instead, "photographs by the author" assumed Edward's presence in Alaska and the Yukon in 1897, and his taking the risks incidental to photographing scenes in an inhospitable area, something Asahel did in fact. The upshot of the fraternal dispute was Asahel's leaving Edward's house. The 1901 city directory lists him at another address and in business as Curtis and Romans, photographers.[40]

Two shots remained to be fired in this consanguineous squabble. An extensive 1988 article by the journalist Sally MacDonald aired an additional reason for Edward's family to dislike Asahel. According to an interview with Edward's biographer Barbara A. Davis, after the family worked "feverishly" to prepare for Asahel's trip, Asahel stayed south of Dawson for a year before arriving at the town and its gold fields. After all, that was where the action was. Edward's children thought their uncle's "behavior in Alaska irresponsible." The claim of irresponsibility took no account of how Asahel was to support himself once in the Klondike, or of his need to innovate to meet conditions, or of any instructions from Edward, or of the slow communication between Seattle and the Yukon, or of the improbability of controlling anyone's movements from a distance of 1,800 miles.[41]

Earlier Asahel's oldest son, Asahel Jr., blamed Edward for the break, but for reasons having nothing to do with the photos. In a 1981 interview he claimed that "Dad ended up taking care of Grandma Curtis [Ellen]." Asahel installed her in "a little house of her own…and took care of Grandma Curtis until she died…Edward S. Curtis never paid anything on the whole thing."[42] Apart from the fact that the "little house" was for many years a residence for Asahel's entire family, his son's reminiscence conflates two very different situations, an earlier and a later one. In the later circumstances Edward did indeed spend less and less time with his family, and more and more time on his Native American photography project. One result was that his wife filed for divorce in 1916.[43]

Edward's later actions, however, did not comport with his earlier support of Ellen and the rest of the family, wholly or partly, from the time he and his father arrived on Puget Sound in 1887. Besides sheltering his mother, Edward gave her money. On February 10, 1890, Ellen transferred ten acres of the Sidney farm to Edward for $1,200, then a substantial sum. Edward sold the property on the same day for a mere $250. Ten days later Ellen sold Edward another ten acres for $1,000. One day later Edward sold the second ten acres for $200, accepting a mortgage note for the entire amount. He had, in other words, given his mother $2,200 for property they both knew would not be worth very much.[44] Asahel Jr.'s condemnation of Edward has, therefore, little more validity than the criticisms of Asahel made by Edward's family. Edward's near abandonment of his family is a matter of record, but it should not cloud his caring for his mother in earlier days.

For all the familial disputation, the brothers were similar in some respects. Both were extremely hard workers, physically robust, mentally acute, and, within the limits of commercial necessity, fiercely independent. Their very similarities mandated their separate destinies, according to those who knew them. Edward was unlikely to take Asahel into full partnership if that meant a genuine sharing of business decisions. Asahel, who barely scraped five feet, six inches, was overshadowed by Edward's six feet or more, but was temperamentally unsuited to being the figurative "little brother." Be that as it may, the break, when it came, did not have to be acrimonious. For the acrimony, Edward, older, presumably more mature, the personal appropriator of his brother's work, yet in control of the situation, must bear the greater responsibility.[45]

So Asahel was on his own, and in a studio partnership with William P. Romans. Marrying Florence Etta Carney on November 29, 1902, was the next major event of his life. The daughter of a Seattle building contractor, she lived until her marriage with her parents, John E. and Ellen J. Carney, and a younger brother, south of the the city's business district. Nothing survives in public records about how or when they met, or about their courtship. The couple was "quietly married" by the Rev. Father Francis Xavier Prefontaine in his library, a location precluding a sumptuous wedding. The arrangement may have been a compromise between Asahel, whose rejection of organized religion would have inclined him to a brief civil ceremony, and his eighteen-year-old bride, who could have wished for a statelier service. She may have chosen Prefontaine's church because it was near her home, or because of the esteem for Prefontaine in Seattle, or because she was a Roman Catholic and Prefontaine was a "secular priest" of that faith. If the religious reason influenced her most, she later altered her theological views, and identified instead with the Pilgrim Congregational Church.[46]

Soon after the wedding Florence and Asahel left Seattle for San Francisco, where Asahel worked as a photoengraver. Possibly the lure of a higher salary than Asahel's income at the Curtis and Romans studio was the deciding factor. The split with Romans was amicable, or else the two would not have renewed their collaboration later. A brief stint in Tacoma followed the San Francisco stay. By 1905 Asahel was back in Seattle. He worked for the *P-I* as a photographer for a year or so, but had rejoined Romans by 1907. This time he was a photographer, not a full partner. Sometime during the next year Romans departed Seattle, leaving Asahel with full responsibility for the studio.[47]

After 1908 Curtis was in charge of the Romans Photographic Company, moving from secretary in 1909 to president-manager in 1910. By 1912 he had formed a partnership with Walter P. Miller, although the pair in 1914 listed themselves as the proprietors of the Romans company. Meanwhile Romans returned to Seattle, his role in the firm unclear, though by 1916 he was gone for good. The next year Asahel and Miller dissolved their partnership, with Asahel retaining the Romans studio. In 1919 the Romans ad in the city directory boasted of "Expert Operators to Cover the Most Difficult Assignments in the Northwest," and of "Lantern Slides Covering the Scenery and Industries of the Northwest and Alaska." Another listing in the same publication declared that "We Photograph Everything."[48]

In 1920 Curtis was not quite forty-six years old. He could look back on an already full life: uprooted at thirteen, bound for the Klondike at twenty-two, and returning at twenty-four, after surviving in an environment that broke many men. The confrontation with Edward, marriage and family, an urban based business combined with a producing farm in eastern Washington, and a growing range of extra-economic interests were all part of the kaleidoscopic experience. Photography and family, though sometimes slighted, continued to anchor his centrifugal life.

Chapter Two

Photographing the Pacific Northwest

NINETEEN-TWENTY was a watershed year for Curtis. He and Lawrence D. Lindsley, of whom more later, incorporated the Asahel Curtis Photo Co., with Curtis taking eighty shares of capital stock at $100 each, and Lindsley five shares at the same figure. Whether the shares were fully paid is not clear. But it made no difference, for Curtis was now his own boss. He had bought out whatever investment Romans retained in the business and folded Lindsley's interest in the Romans company into his own. A 1921 advertisement informed the public that the "Asahel Curtis Photo Co." had become the "Successor to Romans Photo Co." The next year and after the studio was simply the Asahel Curtis Photo Co., although a city directory display ad continued the Romans name for a few years. The Curtis business address remained in the Colman Building where Romans had been since 1909.[1]

While Curtis matured into an independent businessman his family grew. Whitney Asahel was born in 1903 during the California sojourn. Walter followed in 1906, after Asahel and Florence returned to Seattle. Margaret (called Betty) arrived in 1911, and Ellen Jane (called Polly) in 1920. In 1917 the family moved from its long-time house at 1115 36th Avenue to two other addresses successively, and finally to 626 Belmont Avenue North. The move to Belmont occurred about the time of Polly's birth. Curtis lived there for the rest of his life.[2] In 1920 he built a much larger house than the original structure on his ranch near Grandview in eastern Washington (see Chapter Eleven).

Mother Ellen and sister Eva relocated more frequently. They lived at the 36th Avenue address until Asahel, Florence, and young Whitney returned from California, then moved to Edward's house. In 1907 they moved out, remaining at another address until Ellen's death in 1912. Next, Eva bounced among at least three addresses until, in 1917, she moved under Asahel's roof. From then on she worked for Asahel as a bookkeeper and clerk, and in the skilled job of colorist in the days before point-and-shoot color photography. She never married. About 1929 she left Asahel's house to live elsewhere in Seattle. What all this moving meant in terms of family relationships may only be guessed.[3]

Asahel and Florence separated in 1912 and 1913, or perhaps only during 1913, when Florence and Betty, although not, apparently, Whitney and Walter, moved to Portland. They returned to Seattle about the time that Asahel embarked on an elaborate tour of Alaska, an 8,000-mile expedition organized by the Alaska Bureau of the Seattle Chamber of Commerce. Partly a tourist adventure, the trip was also intended to be a serious investigation of the resources and the development needs of the far-flung territory, undertaken in the afterglow of President Woodrow Wilson's signing, in March, legislation mandating a federal railroad for Alaska. The 112 passengers traveled by ocean steamer, rail, and sternwheel riverboats. They spent a month and a week in a journey stretching from Seattle up the Canadian coast, through the Inside Passage, then to Whitehorse, Dawson, Fairbanks and Nome, around the Alaska Peninsula, along the arc of the Alaska coast to the Inside Passage again, and finally south to Seattle. Curtis took many photographs. He could have told and retold the anecdotes gathered in his 8,000-mile transit, but the journey "was most definitely never mentioned," wrote Betty, his older daughter. She "only knew years later," probably when she viewed her father's photos while working for him in the 1930s. The trip may have been off limits for discussion because of a connection to the Florence-Asahel separation, which in turn may have been related to Florence's recurring but remitting illness, possibly a bipolar disorder. Whatever its cause, the separation was not repeated.[4]

In later years Florence was well enough to be active in some voluntary organizations, most prominently the Woman's Century Club, founded by the famed suffragist Carrie Chapman Catt and nine other women in 1891, during Catt's brief residence in Seattle. The Century Club was a well-known social and educational organization featuring teas, dances, picnics, and talks by members and outside experts on issues of the day, its members dividing into departments—Florence's was philosophy and science—depending on personal interests. She joined the club in 1927, when her child-raising duties, except caring for Polly, were largely past.

Though not in leadership positions, she often was listed as assisting at teas or luncheons, giving a paper, or having charge of party decorations. During her active years the club's headquarters were in its graceful red brick building in Seattle's Capitol Hill neighborhood.[5]

Florence's other major social activity involved her children's musical education and, especially, encouraging the two most musically talented, Walter and Polly. If Asahel, whose music appreciation was decidedly undeveloped, took any interest in this aspect of his offsprings' maturation, it went unremarked. Although there is no record of Florence's playing an instrument, she was involved with the Washington State Federation of Music Clubs, patron groups of the Cornish School (now the Cornish College of the Arts), where Walter and Polly studied, and the Music and Art Foundation. She apparently assumed no prominent role in any of these organizations, but was a club leader in the Camp Fire Girls at the time Polly would have been interested in the organization.[6]

Meanwhile, the client list of the Asahel Curtis Photo Co. grew. Given Curtis's devotion to Mount Rainier, it is not surprising that the National Park Service was a beneficiary, at times, of his photographs at no cost. The Bureau of Public Roads and the Bureau of Reclamation also commissioned his work. Yet he was diffident about seeking government jobs, whether directly or indirectly through private parties lobbying Congress or government agencies. He conceded that the photo albums he prepared for members of Congress showing irrigation farming were effective, "but I am reluctant to appear to be seeking to use this as a method of securing business for my own firm."[7]

No such qualms prevented him from garnering commissions from private companies or groups. He was for a time an official photographer for the Northern Pacific Railroad. The Great Northern and the Chicago, St. Paul & Pacific (the Milwaukee Road) also employed him. He took photographs for the chambers of commerce of Seattle, Spokane, and Yakima. National customers included the American Steel and Wire Company and the Portland Cement Association. Among the many local or Seattle area clients were the Apex Fish Company, the Associated Terminals Company, the Lewis Construction Company (developer of Seattle's vast Jackson Street regrade), the Puget Sound Navigation Company, and the Thomas Investment Company.[8]

The range of his photographs is astonishing, especially considering his concentration on the Pacific Northwest. People in all their varied public and semi-public activities formed a significant part of his personal collection. He photographed them working, relaxing, playing games, marching or riding in parades, and enjoying the outdoors. He created portraits of leading Seattleites and of people less prominent, as well as group shots. He photographed people and their animals ranching and farming, especially on or near his own ranch. His scenic photos included hundreds of Mount Rainier, of fellow climbers assaulting its slopes, and of views around Mount Rainier National Park, not to mention other mountains; city, state, and national parks; and national forests. Transportation in all its forms fascinated him, an absorption which helped to fill the coffers of the Asahel Curtis Photo Co. when he translated it to glass plates or film. Horse-drawn wagons and carriages, trolleys, trucks, automobiles and airplanes, boats, ships, warships, and trains, all are preserved in his huge collections. Outdoorsman though he was, thousands of his pictures are resolutely urban. There are street scenes in daylight and dark, churches, schools, public buildings, public works, and office buildings. He photographed the Alaska-Yukon-Pacific Exposition held in Seattle on the University of Washington campus during the summer and autumn of 1909, even though he was not the official photographer.[9] He wrote dozens of newspaper and magazine articles, many of them illustrated with his own work.

Much of Curtis's oeuvre related to his role as an inveterate booster for Washington State and the larger Pacific Northwest. As a booster he enthusiastically publicized the scenic riches, natural resources, and industrial development of Washington. He did so through his Seattle Chamber of Commerce activity, his fight for the good roads necessary to bring tourists and new residents into the state, his struggle to expand irrigation and reclamation, and, most of all, his effort to open Mount Rainier National Park to the tourist and recreationist. Early national recognition came in 1917, when he traveled as far west as Chicago, displaying twenty-six large, hand-colored photographs of highway scenes between Seattle and the Illinois metropolis. He undertook the trip on behalf of the National Parks Highway Association, and later exhibited the photos at a Seattle department store. Capitalizing on a suggestion from the director of extension services at Washington State College, from 1925 onward he gave illustrated lectures to schools, clubs, and other audiences around the state, focusing on the scenic beauties of the commonwealth. He probably arranged most of these slide shows to coincide with travels to visit his Yakima Valley ranch or on photo assignments.[10]

A brief collaboration with a friend produced the Marsh-Curtis Know Your State Bureau located in Curtis's studio. This organization promised to extend his shows by inculcating civic pride through illustrated lectures about the scenic, industrial, and natural resource attractions of Washington. Curtis could not devote much time to it because he was involved with so many other activities, and because he found little interest in an overt "know your state" idea as opposed to regional enthusiasms. Soon he was using Marsh-Curtis letterheads for carbon copies of his correspondence. Despite the collapse of Marsh-Curtis, Curtis's own letterhead carried at the bottom "Know Your Own State" in capital letters, and beneath that motto, "Boosters Live Better, Feel Better, Fight Harder than any other tribe on earth." The reverse of each letterhead featured Curtis photos and the slogan, "Washington: Nature's Paradise—Man's Opportunity."[11]

Curtis's boosting appears intellectually shallow, and to an extent it was a puerile activity, but he was a true believer in publicity, unembarrassed by strident appeals to Pacific Northwest exceptionalism. He was convinced of their merit, and of their influence in persuading Washingtonians and outsiders of the extraordinary beauty and advantages of the state and region. Statements that seem superficial or even disconcerting to later generations were perfectly acceptable to him. Nor should anyone suppose that boosterism, especially the species devoted to snagging tourists, disappeared with the introduction of more elevated tastes. A glance at later city, state, or regional advertising will disclose that sloganeering has not vanished from the booster lexicon, although the language may suggest a softer, sophisticated refinement.[12]

Three more observations about Curtis's boosterism are in order. The first is that its results were modest. Certainly they are not quantifiable. In 1927 he wrote about his 1924 "photographic survey of the Olympic Peninsula." One outcome of the survey was "150 slides…shown to more than 250 audiences" statewide, "with the result of a very material increase in the travel to the Olympic Peninsula and the development of a series of Chalets for…tourists."[13] The slide shows may have had some impact, but it is impossible to disentangle their influence from rising 1920s prosperity, improving roads and cars, and growing Forest Service interest in promoting tourism in what was then a national forest. In 1929 he urged a correspondent to print a map on the reverse of all stationery, writing about how such a map increased travel to Mount Rainier National Park.[14] The map did not prevent a decline in park visits during the Great Depression.

The second observation involves Curtis's disinterest in advancing his own business directly through boosterism. If boosting succeeded and the Pacific Northwest prospered, his studio's business could expand, to be sure. His concern was nonetheless limited to, as the phrase goes, the rising tide that lifts all boats. "I have a strong interest in the Yakima Valley aside from my own holdings and would like to see the country get more publicity," he wrote in 1912.[15] He deplored sectionalism and city rivalries. He detested the 1920s struggle over changing the name of Mount Rainier to Mount Tacoma for several reasons (see Chapter Seven), but one of them was the damage the conflict had on the united effort to draw tourists into the park. He despised the "obstacles placed in the way of" park improvement "through the jealousies of the people of our State," and the resulting "untold discord in the State of Washington."[16] Harmony and cooperation in advancing Washington's interests were fragile partly "because Washington appears to be community minded instead of state minded."[17]

Curtis's desire for statewide harmony clashed at times with other principles such as scenic preservation and conservation, resource exploitation, and industrial development. When that happened, he had to compromise his convictions. His accommodations to reality and the resulting disharmony were not always made gracefully, and they were subject to misunderstanding and distortion, as later chapters will show. To illustrate his problem with one example, he firmly believed in the economic and social contributions of good paved highways, and equally in the timber industry's participation in economic and social progress. Alas for Curtis's emotional comfort, truck damage, including logging truck damage to highways built for much lighter cars, was a serious issue in the 1920s. Curtis had to choose. He came down on the side of lawful load limits and highway protection. "I believe that it is an economic loss to handle logging trucks on our state highways and personally would like to see a law prohibiting the practice." Present laws were, however, good enough, and if "rigidly enforced" would make logging trucks "unprofitable."[18]

Curtis's belief in championing the entire state leads to the third observation about his boosterism. It should not be confused with promotionalism, even though the line between boosting and promoting could be blurred. Curtis, the state and regional booster, was an enthusiastic publicizer of his chosen area while the promoter was directly concerned with making money. In this example the promoter organizes a company and sells the company's product, whether the product is land, buildings, or anything else, to the public with the motive of realizing a profit. A promoter such as a chamber of commerce or a trade association operates as a promotional collective of the interests giving it financial support, with the purpose of helping each individual interest to prosper. This booster/promoter distinction is critical to understanding Curtis's role as a photographer. Many companies paid Curtis to photograph their works or holdings as part of their promotional activity. One example is his photographic contribution to the promotional brochures of the irrigation companies in the Yakima Valley (see Chapter Eleven). In these cases Curtis was a commercial photographer for customers who valued his work and incorporated his pictures in their promotions. He was the vice president of the Smyser Display Service, a Tacoma firm, winner of the state contract to create the Washington exhibits at the 1939 world's fairs in New York and San Francisco, but, the display company aside, there is no evidence of an investment in any other property beyond Seattle houses and his ranch in the Yakima Valley. In short, he engaged in no direct promoting save the modest local advertising of the Asahel Curtis Photo Co.[19]

Understanding the limits of Curtis's booster role is essential to assessing the criticism of his work arising from later preoccupations with the effects of explosive population growth, massive industrial development, sprawling residential expansion, and rampant resource consumption. The backlash against unbridled growth began in the late 1950s and gathered momentum in the following decades, resulting in vastly increased government controls, no-growth policies, and extreme efforts to shift energy consumption from nonrenewable to renewable resources. The reaction against growth, development, and use was reflected in thinking about past champions of growth, development, and use, Curtis included.

The earliest study of Curtis photographs (1973) in the post-1950s era is emblematic of the harsh criticisms to follow. In the Forward to *The Asahel Curtis Sampler,* Murray Morgan damned its subject with faint praise. Morgan, an accomplished popular historian, wrote of Curtis that he "loved the outdoors, but in promoter-fashion: let's develop it." The potential lack of "enough outdoors to go round was a concept foreign to Curtis and most of his contemporaries." According to Morgan, Curtis's photos, though often poorly composed and developed, are nice to have, not for their quality but for their depictions of Pacific Northwest growth. David Sucher, the editor, wrote in a similar vein. He found Curtis's correspondence devoid of "any mention of art or culture or politics" as though his interest in Mount Rainier National Park, for example, did not result in art photos, did not concern the cultural value of recreation, and did not involve the often cutthroat politics of National Park Service appropriations. Curtis's good roads work was practically all political, and despite his rejection of exotic plants for roadside decoration (see Chapter Five), Curtis was "no environmentalist." His vision of the future was "a straight-line projection of the present," though in 1910 he was fascinated by the underdeveloped, new-fangled airplane, well before World War I raised international awareness of its utility in war and peace.[20]

A second popular historian, Archie Satterfield, opened another window to the thinking about boosters in the later twentieth century. An unpublished segment of what became a book on Curtis's Seattle photos justified early urban boosting as a survival mechanism, declaring Curtis to be "an undisputed master at the art of boosting." A few years after Curtis died, however, "boosterism had served its purpose…and the younger people…had more cosmopolitan issues in mind. A growing number of people consider the city-growth advocates potentially dangerous to the very city they were trying to serve because they could not see the inherent dangers in too much growth." But according to Satterfield's critique, the "inherent dangers" were not evident during Curtis's lifetime. Therefore the criticism applies only to post-Curtis boosters and not to Curtis himself.[21]

Curtis should not be excused from criticism, as when he wrote intemperate letters during the 1920s fight over changing the name of Mount Rainier to Mount Tacoma. By the same token he should not be criticized for attempts to overcome the relative isolation of the Pacific Northwest by encouraging immigration, investing, and tourism. Curtis the booster was of course aligned with the promoters, taking their money in return for his photos. At the same time he was uninterested in promotion for its own sake, because becoming a promoter would have required an investment in something to promote. Turning to the specific activity of promotion would have compromised his broad vision of state and regional prosperity based on the harmonious participation of all interested parties.

Photography formed the core of Curtis's work despite his involvement in activism, boosterism, farming, and family. His Klondike photos invite comparison with those of other gold rush picture takers whose work was vivid. Curtis recorded nothing as sensational as Eric A. Hegg's iconic photos of Klondikers struggling up a thirty-degree slope to the summit of Chilkoot Pass, forming a gritty black garland over the stark white snow. Hegg opened a studio in Dawson as Curtis did not, and so preserved interior scenes in the town.[22] Frank La Roche photographed prostitutes wading the Dyea River on their way to Dawson, wearing the heavy travel clothing of the era. His less well known but more winsome photo shows them in camp on the British Columbia portion of the trail, some of them smiling and most of them wearing the boots, knickers, and heavy shirts of camp attire. In a bow to Victorian sensibilities La Roche labeled them "actresses," which they were, in a sense. His picture of boat building on Lake Lindeman in 1897 is a masterpiece, almost stereographic in its sharp focus from foreground to distant hills.[23] Frank H. Nowell's photographs of the gold rush slight the Klondike itself but provide an important record of Juneau, Fairbanks, Nome, and other towns. His studies of Natives were made without any ethnographic pretense but are valuable documents nevertheless.[24] Curtis's work is limited to Skagway, the trails, Dawson, and the gold fields.

Curtis rates well, however, when his circumstances and some of his best photos are considered. All of his rivals were older and, except possibly Nowell, had more camera experience. Hegg, who opened his first studio at fifteen, was born six years before Curtis. Nowell was ten years older. La Roche was Curtis's senior by twenty-one years. All spent more time in the Alaska-Canada area, Nowell arriving first in 1886.[25] Many of Curtis's shots compare favorably with Hegg's, especially those of the streets of Dawson, the Whitehorse Rapids, individual Klondikers, and miners at work.[26] Some of Curtis's gold rush pictures are masterful, such as his beautifully composed moonlight shot of two stampeders dragging their sleds along a trail. The scene is carefully staged during a warm spring night. The partly bare rock all around, the trees almost free of snow, the stratified, compacted snow exposed at the left of the stampeders, and especially the open water in the foreground, all confirm a warm springtime. It is warm enough for the men pulling their sleds to stand stock still for the moonlight exposure. A lightening at the right of the photo and the fall of shadows away from the light suggest an early Northland daybreak arriving to shorten the exposure time. Indeed the probability is that the scene is not on any main trail but on a little-used side trail or on a "trail" leading nowhere but especially prepared for the purpose.[27] Curtis's portrait of Charles Ainsworth, his mining partner, playing solitaire is a study in lighting and composition. Most of the light falls on Ainsworth, bearded, his hair tousled, his thick shirt rumpled. He is concentrating on the dog-eared cards. The background reveals their rough miner's cabin, with what looks like an oilskin slicker carelessly tossed, plus a bottle and miscellaneous cans. No viewer would confuse the setting with a gentlemen's clubroom.[28]

Back in Seattle, Curtis snapped scores of urban scenes, as did his contemporary, Arthur Churchill Warner. Both took workaday photos of buildings as they probably were paid to do, but both men also recorded vivid streetscapes. Another Curtis contemporary, Darius "Dee" Kinsey (1886-1945), focused on the logging industry in all its aspects, though he also photographed his family, groups, Native Americans, and Mount Rainier. Curtis's logging photos, while not as numerous, are as dynamic.[29] In sum, his photographic record compares favorably with his professional contemporaries', matching or sometimes exceeding much of their best work.

Almost every mention of Asahel also mentions Edward, if only to dramatize the fraternal fracture driving them apart. Edward's mature photography is markedly different from Asahel's, so what draws them together, Alexander I. Olson argues, is marketing the western appeal of a sentimentalized, contrived "heritage" involving Indians. Olson centers his argument on Asahel's photographing a 1925 Spokane, Washington, Chamber of Commerce sponsored "Northwest Indian Congress." Nothing permanent developed from this congress but the Northern Pacific Railroad and others used some of Asahel's photos in their promotional work. Olson finds Asahel walking the low road of heritage production, snapping staged photographs of Indians in traditional dress, sometimes against contemporary Spokane backgrounds. These he surrendered to his paymasters, who failed to bring their Native American clients into the sort of networking and infrastructure development necessary to a permanent exhibition. Edward, on the other hand, took the high road of heritage manufacture, implying that his carefully contrived photographs of individual Indians and their group ceremonies, as well as his lectures and his elaborate fieldwork, had no commercial object. Edward's activity, in reality, was as monetary as Asahel's. It was even more immersed in the western craze for packaging and selling a regional product to midwestern and eastern tourists.[30]

Olson's approach is a corrective to presentations of Asahel and Edward as dramatically different chroniclers of the world around them, although a review of the extensive work on Edward published before Olson's article would arrive at a similar conclusion.[31] Mick Gidley's study, first appearing in 1998, demonstrated Edward's need to mine every possible source for money. Edward's elaborate photographic, ethnographic, and anthropological field work involved a varied, shifting team of specialists, including interpreters. Its result, the monumental, twenty-volume *The North American Indian,* appeared between 1907 and 1930. The stupefyingly wealthy banker J. P. Morgan supported his effort in part, and President Theodore Roosevelt endorsed it. Edward sincerely believed Indians to be a "vanishing race" in the sense of the declining numbers of Native Americans, and in the meaning of their absorption into the larger culture. He wished to preserve a record of them and their accoutrements in the late precontact stage. Thus his extensive use of precontact modes of dress, decoration, and other paraphernalia to authenticate his subjects, and his elaborate employment of photo techniques such as cropping and dodging to eliminate contemporary intrusions on his images. Developing his pictures in sepia tone gave them a quality of great age as well as timelessness. His portraits, especially, are striking, quite apart from the debate over their contributions to anthropology and ethnography.[32]

The flood of writing on Edward's work negates the need for more extensive commentary, except to note a few more comparisons between Asahel and Edward. Asahel took many photos of Native Americans, but accomplishing contemporary tasks. His pictures of the Neah Bay (northwest Washington) Makah at their whaling depict the Makah in contemporary clothing, going about the dangerous, disorderly business of catching, landing, and carving up a whale. His photographs of Indian hop pickers are posed, for a payment, showing, for example, a man using a long pole to pull down a hop vine while the women wait to begin picking. The hop workers are garbed as anyone else would be in similar agricultural work. Asahel showed Indians in their contemporary, everyday activities; therefore his photos are more honestly authentic than Edward's carefully contrived scenes. Valuable as Edward's work is, Asahel made the superior record of early twentieth-century Native American life.[33]

Before and after he established his own studio in 1920 Asahel never intended his photography to be an end in itself. Rather, it was a way to support his family and his other activities. At times he curtailed other commitments to maintain his studio (see Chapter Six) but usually he tried to balance his running the business with his ranch and other interests, all the while understanding how his involvement elsewhere impacted his firm. "It is really a very expensive luxury for me to get away from the office at extended periods," he wrote in 1922. His business was "to a considerable extent personal and unless I am here, certain features bog down in spite of anything that my office force can do."[34] The office force varied from eight or nine in the prosperous twenties to three or four regulars in the hungry thirties. One or two reliable helpers were called in, part time, when the work was heavy.[35]

The studio enjoyed a good income despite Curtis's absences. Sales and interest totaled $21,109 in 1925 (about $283,500 today). The solid years of the twenties contrasted with $7,967 in 1932 (about $137,000 today) and $7,849 in 1935. Most years showed a small "surplus" after deductions for all expenses, surpluses apparently distributed to employees. The studio survived but the halved income hurt. "I am as broke as they make them," Curtis admitted in 1933, and as he was between secretaries, "have to hammer out my letters myself."[36]

Several accounts of the studio's operations survive, most extensively recounted in the diary of Lawrence Lindsley, a scion of the founding Denny family of Seattle. It is not a record favorable to Curtis. Lindsley was a superb outdoorsman and photographer, but he was also a chronic debtor, a hypochondriac, and a confirmed whiner and complainer. He found little solace in whatever he was doing at the moment and instead wished for something else. Early in 1914 he was working with Curtis in the Romans studio, but anxious to be "home" at his parents' property on Lake Chelan, east and north of Seattle across the Cascades. He returned to Lake Chelan in March and became engaged, but by February 1916 he wanted "to get down to Seattle for a while for a change," to work with Curtis or at some other job. Eventually, in mid-December, Curtis took him on. Lindsley promised himself to "try my best" and to "make good." By early January 1917, however, he was dissatisfied with Zerby Strong, another employee, and disgusted with what he saw as "any old way" to get the work done. For the next fourteen years he stayed on at the studio, compiling a litany of criticism. Curtis would not pay attention to business, the work areas were dirty and disorganized, another fellow worker, not Strong, was a "snake in the grass," and "Curtis don't seem to have any head for running a business at all," because Curtis would not give him detailed instructions. When, in 1927, Curtis was away for two weeks on a Columbia River Basin tour, Lindsley fumed. "This trying to save the world don't pay," he told his diary. In December he decided that "Curtis is only a small man and in a deep rut—Has been running on his brother Ed Curtis' reputation and national fame."[37]

Lindsley's last comment, one he repeated in a later interview, would have amazed Curtis or anyone else who knew of the brothers' profound estrangement. Asahel made his own way in the world without any significant help from Edward following their separation. Nor could he have survived on his brother's "reputation and national fame," for neither reputation nor fame were Edward's by the late 1920s. Edward continued to work furiously but his nationwide recognition peaked years before. The revived appreciation for his oeuvre dates from the 1960s. Yet Lindsley's diary reveals the continuing, if tenuous relationship between the Curtis brothers. On May 13, 1918, he wrote: "Guess Mrs. Strong is going to the Ed Curtis Studio...Says she will get $45 or $50 per week. Wonder what Curtis will do for another color artist." Within two years Strong returned because Lindsley recorded an April 1920 "scrap" with her over the use of a tray. It is difficult to imagine Strong not being a conduit between the brothers. Lindsley also noted working on photos for Edward's "Vanishing Race" series, a claim he repeated in the interview. Asahel's business records do not support the assertions but the collaboration could have happened.[38]

In other respects Lindsley's diary is revealing of Curtis and of himself, too, in ways he did not intend. True to form, Lindsley soon wanted out of Curtis's studio and to be working elsewhere or in his own business. He could hardly have concealed such a strong dislike for the office, or for Curtis himself, from his boss. At least as early as 1919 Lindsley was operating a studio in his house, activity that could be seen as diverting income from Curtis. In the same year he tried to get on at Webster & Stevens, major Seattle photographers, but was unsuccessful. His attempt could have been brought to Curtis's notice. Any of these actions or attitudes would have been grounds for dismissal had Curtis chosen to make an issue of them. He did not. Instead he gave Lindsley two weeks off and money for a week's vacation when he married in 1918 and a week off when his wife died in 1920, generous compensation by the standards of the day. Lindsley was a gifted photographer and doubtless a skilled enlarger, his principal job, two reasons for Curtis to retain him despite his poor attitude.[39]

Curtis's younger daughter Polly echoed one of Lindsley's criticisms in a 1988 interview. Her father, she said, "liked to think of himself as a shrewd businessman, but…I think he had no knack for business." There are problems with this view of Curtis. Neither Lindsley nor Polly understood the function of the studio—to provide income for the Curtis family and for Curtis's other pursuits. Polly was not close to her father, indeed she "resented that Dad could find so much time to spend on so many projects," turning his children inward and making them ambitious "for ourselves." In fact, a 1908 letter from Curtis to Romans showed Curtis to be conversant with income, expenses, collections, and the flow of work in the Romans studio. His 1922 letter at the head of this discussion reveals his understanding of the cost to his business of his varied interests. His daughter Betty, who worked for him in the 1930s, knew "for sure" that his stay in Washington, DC, to testify for a smaller Olympic National Park than the one adopted (see Chapter Ten) "cost him lots of money thru his absence." Her father "lost money on the whole affair," but capitalizing financially on his concern "was not his purpose."[40]

An interview with three women who worked for Curtis from 1920 indirectly refutes the view of him as an indifferent businessman. Instead he emerges as a man who expected his employees to take the initiative in his absence. When he returned after a month in the Olympics in 1924 his "efficient office staff" had things in order and "there was money in the bank." Curtis's response was: "Well! Maybe I should go away more often." The remark was as close to humor as Curtis came with his employees, for he did not lighten the office atmosphere by joking with them. In a day when first-naming was restricted to people who were well acquainted peers, and formal titles such as "mister" followed by the last name where common in office situations, Curtis used patronymics in the office, but without the softening of the title. His practice applied to women as well as men, thus Harriett Holmberg was "Holmberg." There is no record of what he called his sister Eva or daughter Betty while at work. Even so there were lighter moments at the studio. The women found it easy to climb from their workplace on the top floor of the Colman Building to the roof to have lunch on nice days. They had nicknames among themselves.[41] Even Lindsley remembered jokes about an illustrated Klondike lecture of Curtis's, titled "The Spell of the Yukon," after the Robert Service poem. Curtis so frequently spoke of it, revised it, and required the staff to make new slides for it, "that we called it 'Smell of the Yukon.'"[42] Though reserved while on duty at the office, Curtis was not standoffish. Friends found "Asahel" too difficult a first name, so called him "Ace" or some other nickname. Some of them gathered at his cluttered studio on Saturday afternoons whenever he was in town. There he held a "sort of open house" where "old friends dropped in to hear his stories of Northwest characters" and of the "Klondike," stories that doubtless lost nothing in the retelling.[43]

Curtis enjoyed a well-developed sense of humor, as the last anecdote illustrates, but his sense of humor was not always at the forefront. One biographer described him as "dour." True, Curtis was negative or hostile in some situations. He had no use for organized religion and its ceremonies, including burials. His father filled his childhood with Sunday church, bans on play or work on the Sabbath, Wednesday prayer meetings, prohibitions against frivolous amusements such as card playing and dancing, and hour-long prayers at evening meals—at least to young Asahel, the prayers seemed to continue for an hour. One friend remembered that Curtis had no "small talk" and would "clam right up" when he was uninterested in a subject. He could be sarcastic and opinionated. Women journalists who began an interview by praising his photographic artistry found their interviews curtailed. They correctly described Curtis as an artist, for some of his photographs are artistic in their composition and staging, and any photograph is a work of art in the sense that it is a representation of its object, not of reality. Despite the artistry he displayed, Curtis rejected any description of his work as artistic.[44]

Some appearances aside, Curtis's often wry sense of humor was on display when circumstances provoked it. A lumber company in Yakima, angling for a complimentary photograph, sent him a letter declaring "one of his timber photos" to be "the greatest they had ever seen." If he would send a free copy to the company, the firm "would display it prominently" and suggest that interested people purchase additional copies. Curtis "replied that he would be glad to send them a free copy of the photograph if they would send him lumber to build a new house, in front of which he would put a sign telling people where the lumber came from and recommending they buy their lumber there." Soon he received a check for the photograph. On another occasion he became dissatisfied with the narrative ability of a partner on one of his slide show circuits, so he took over the lecture. "But I want to be in the show," the demoted partner insisted. Later Curtis found a photograph of a totem pole, replaced a carving on the pole with the unhappy partner's face, made a slide of the concoction, and had Eva color it. When he next saw the ex-partner, he showed him the slide and declared, "now you're in the show."[45]

Late in 1931 he parodied some of the work relief fantasies of the Great Depression. One of them, a scheme that surfaced in several versions, was a plan to build a tunnel through the Cascade Mountains connecting eastern and western Washington at an elevation low enough to avoid heavy winter snows. Other notions involved canals across the state. Curtis offered a plan to divert the Columbia River to Puget Sound through a series of dams and a tunnel carved through the Cascades, making Seattle's rival "Portland the inland city the Creator intended it to be." Huge fans in the tunnel would blow moist air from Puget Sound to eastern Washington, obviating the need for irrigation. The fan arrangement would have to be self-supporting, so if farmers refused to pay, the fans could be reversed, a bitter comment on the inability of many farmers on reclaimed land, Curtis included, to pay their bills during the Great Depression. In the event of the fans being reversed, they would send hot air from eastern Washington to Seattle. "Seattle, as Mr. Curtis brightly remarks, is used to hot air," ran one comment on his spoof. Spoil from the tunnel excavation would serve two purposes, Curtis declared. First, it would fill Puget Sound south of Seattle, ending Tacoma's role as a rival port. Second, the remaining earth would create a mountain near Tacoma, "and Tacoma could name it anything she wanted to," a reference to the recent Mount Rainier naming dispute.[46]

In 1932 Curtis focused on the deteriorating international scene, writing a mock-serious essay advocating "a League for the promotion of war." His imagined league developed from his belief that peace organizations had failed to prevent war, so "a league for war" might produce peace. He proposed turning all war preparations over to the league, which would require nations to pay for wars in advance, enforce rules of warfare, and otherwise put war on a "business basis." Politicians, prisoners, and other people "of least use to the nations" would staff the fighting forces, avoiding "the loss of the young men." He knew that no country would surrender its sovereignty to such a league, so his purpose was to suggest the expense, futility, and destructiveness of war by inverting the standard antiwar argument.[47]

Curtis's spoofing involved his family in at least one practical joke. On that occasion he brought home an unusual cut of meat which he declared could be snake or eel. The family knew him to try almost any food, as befitted a hiker and camper. Florence prepared the meat but the family dined only on potatoes and other vegetables. "That's the best loin of pork I ever had," he announced to the astonished group at the end of the meal. There was not a lot of joking and kidding in the family, but its other members did tease Curtis about his attire. His daughter Betty recalled how his camping jacket was fitted with pouch pockets for camera equipment, cigars, a magazine, and other items. Back home in Seattle, he expected his business suits to perform the same function, with the result that they soon were shapeless. Those prematurely aged suits were a source of mirth.[48] Betty, and probably the rest of the family, hugely enjoyed a political joke played on Curtis. A "staunch" Republican, he once received a welcome from a local Democratic committee. Someone, he never discovered the culprit, volunteered him to serve as a Democratic committeeman. The joke "was a darned good" one, Betty remembered.[49]

For all that, Curtis was a shy, private man whose humorous comments generally were mild, or who found humor in "odd things," that were much less than rollicking jokes but nonetheless caused him to "chuckle over them for days." He could write scorching letters but rarely criticized anyone verbally. When he did, "poor prune" was the strongest condemnation he made of anyone. One such "poor prune" was Dave Beck, by 1930 the powerful head of a Teamsters' Union local and an organizer for the international headquarters. As Betty told the story, Beck refused to pay $150 (about $2,100 today) for some gold tone prints, declaring them to be Curtis's contribution to the union. That Curtis would make such a contribution voluntarily was beyond belief. He could not collect the bill, so agreed to let Betty try. Betty took a sack lunch to the union offices and sat on a bench for two hours. Then a Beck assistant emerged, handed her the "damned money" and ordered her to "get out."[50]

Curtis was an avid reader, not surprising for one who could withdraw into himself despite his demanding schedule. He read the naturalists Henry David Thoreau and John Muir, as well as every Charles Dickens novel. He often quoted Shakespeare and "knew the Bible almost by heart," a credit to his retentive memory and a holdover from his early days of enforced religious practice. His reading explains in part his success as an article writer, because his grammar school education ended either before or soon after he arrived at Port Orchard.[51]

Curtis's other tastes were less well developed. His favorite song was "Old Black Joe," the melancholy Stephen Foster melody about an ancient African-American man, presumably a slave, whose friends are dead and who nears the end of his own life with sadness. Robert Service was his favorite poet, though Service's poetry about the Far North received mixed reviews. "He smoked constantly," Betty remembered, a habit that may have contributed to his fatal heart attack. He puffed "Between the Acts" cigars, small cigarillo types designed for gentlemen to smoke between acts at a theater, or on any occasion when there was no time for a large cigar. Lindsley remembered him scattering ashes and spent cigars on the floor of the studio workroom. Despite his aversion to all things funerary the family did hold a service for him, and, Betty said, "I really felt that he looked entirely wrong at his funeral because" no little cigar was in his mouth.[52]

By the later 1930s Curtis was a business success whose work gained him a regional reputation despite his secondary interest in operating a photography studio. His children were grown or growing to adulthood. The

Great Depression and other issues such as the increasing sentiment for an Olympic National Park frustrated his hopes for the development of Washington and the Pacific Northwest. His life was not all frustration, however, partly because of his leading role in advancing good roads and developing Mount Rainier National Park.

Mount Rainier from Lake Washington

About 1903 Curtis photographed Mount Rainier from a point well above the practically undeveloped shore of Lake Washington. *Washington State Historical Society, Tacoma, 1943.42.4504.3.*

Eleven years later Curtis returned to the lake. By then Lake Washington Boulevard skirted the shore as seen in this photo taken a few feet from the roadway at a different vantage. *University of Washington Libraries, Special Collections, A. Curtis, 31258.*

Four Men Who Worked with Curtis

Owen A. Tomlinson, the superintendent of Mount Rainier National Park from 1923 to 1941, consulted closely with Curtis on park development issues. Their friendship cooled when Curtis published a 1936 article opposing the creation of the Olympic National Park. In this Lawrence Lindsley photo, Tomlinson looks askance at a "camp robber" perched on the barrel of his revolver. *University of Washington Libraries, Special Collections, 28255.*

J. J. "Jack" Underwood was the most effective lobbyist for the Rainier National Park Advisory Board, which Curtis headed for many years. Underwood, gregarious and persistent, worked well with Interior Department officials and members of Congress to increase appropriations for Mount Rainier National Park. *Washington State Historical Society, Tacoma, 1943.42.53360.*

A first-class photographer in his own right, Lawrence "Lawrie" Lindsley was a rugged outdoorsman, as his attire in this trick photo of two Lindsleys suggests. Lindsley worked for Curtis for a decade and a half, filling his diary with criticisms of Curtis's business methods. Nevertheless, Curtis valued his ability and kept him on until the Great Depression forced staff cutbacks. *University of Washington Libraries, Special Collections, 33494.*

Curtis snapped this photo of Governor Roland Hartley (center) in September 1927, when relations between the two were less than cordial because of Hartley's opposition to highway funding that Curtis supported. The diminutive governor (1925-1933) left his hat on to minimize the height differences between himself and the two men flanking him, while his frosty gaze reflects his unhappiness with Curtis. Hartley and Curtis reconciled after the Great Depression changed Hartley's mind about state spending on highway construction and the jobs it produced. *Washington State Historical Society, Tacoma, 1943.42.52462.*

Visual Joking

Curtis was not noted for his robust humor, yet a few of his photographs betray his subdued sense of fun. He took this joke photo of fifteen-year-old Eleanor Chittenden in 1907 while he headed the Mountaineers expedition to climb Mount Olympus. Eleanor stands on a rock in the Elwha River on the Olympic Peninsula, holding a steelhead trout in her right hand while her left hand grasps a fishing pole. The pole lacks a line and reel, so could not have been used to catch the fish. Eleanor's expression indicates that she was in on the joke and helped to set it up. Her presence on the expedition proves how much her father, the renowned author and engineer, Hiram M. Chittenden, trusted Curtis with his daughter's safety. *Washington State Historical Society, Tacoma, 1943.42.8396.*

The settings for some Curtis joke photos were at or near his irrigated ranch west of Grandview in parched eastern Washington. Betty, Curtis's older daughter, here about four years old, appears to wield a hatchet to nail the top slat on a box from the cooperative Yakima Valley Fruit Growers Association. Although Betty's dirty smock lends an air of authenticity to the scene, her knowing smile suggests that a four-year-old nailing fruit boxes with a sharp-bladed hatchet in a box assembly area was not an activity to be taken seriously. *Washington State Historical Society, Tacoma, 1943.42.33478.*

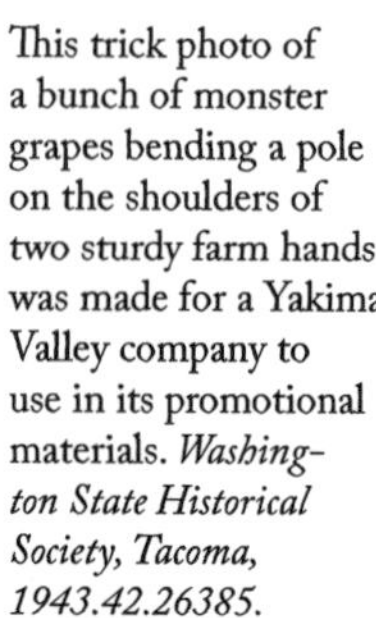

This trick photo of a bunch of monster grapes bending a pole on the shoulders of two sturdy farm hands was made for a Yakima Valley company to use in its promotional materials. *Washington State Historical Society, Tacoma, 1943.42.26385.*

A reluctant horse appears to resist doing its part to break up the soil of the Curtis farm though Curtis himself urges it on from behind, while an employee or neighbor named Norton tugs on the bridle in front. The horse in this April 1918 photo is not resisting anything but is standing still, waiting for the foolishness to end. *Washington State Historical Society, Tacoma, 1943.42.37656.*

In October 1925 Curtis spoofed the Western Euro American obsession with objectifying Native Americans with this study of Mike Littledog, probably a Yakima Indian youngster. The event was a three-day carnival in Spokane, the Northwest Indian Congress. Curtis took many photographs of Native Americans in traditional clothing for the event's sponsors, usually in the incongruous settings of Spokane. The sponsors included the Northern Pacific Railway. In this photo young Mike wears a feathered bonnet many sizes too large for him, while in the background a group of adults is absorbed in a Congress happening and almost pointedly uninterested in Curtis and the child. Probably, Mike posed for a fee paid either to him or his parents. Curtis was willing to photograph almost anything anyone paid him to photograph, but he was more concerned with development activities he considered legitimate, such as providing first-class highways and amenities in Mount Rainier National Park. For more information on the Congress, see the article by Alexander I. Olson listed in the bibliography. *Museum of History and Industry, Seattle, Burlington Northern Photograph Album, 1984.84.7.79, 49708.*

The Far North

In this detail of an 1897 photo Curtis looks every inch the old sourdough trail hand, which he was by the spring of that year. He probably assembled the box camera he carries in his left hand. *Washington State Historical Society, Tacoma, 1943.42.46019.*

During a springtime, probably in 1898 or 1899, Curtis pointed his camera at the setting moon to capture a carefully staged scene illuminated by the moon and the rising sun. Note the bright sunlight at the upper right of the photograph and the fall of shadows away from the light. Open water in the foreground, exposed, layered snow to the left of the human figures, and the lack of winter snow on the surrounding trees all indicate a Far North spring. The men are not packing on a well-worn trail but are standing stock still in lightweight clothing. This is one of Curtis's most skillfully composed early photos. *Washington State Historical Society, Tacoma, 1943.42.46036.*

On a relatively warm day, probably in Dawson City, Yukon Territory, Curtis (left) examines a fish. *Washington State Historical Society, Tacoma, 1943.42.20120.18.*

Curtis's portrait of partner Charles Ainsworth relies on natural light from the window, a rarity in Yukon mining cabins. Ainsworth is a study in concentration on his game of solitaire and its well-thumbed, dog eared cards. Ainsworth's rumpled shirt, the clothing heaped to his left, and the cans and bottles randomly set in front of him are consistent with the careless attitude Curtis maintained in his later life toward matters he considered unimportant. *Washington State Historical Society, Tacoma, 1943.42.46171.*

Chapter Three

The Good Roads Movement: Growing the Washington State Good Roads Association

Curtis joined the good roads movement by 1911 at the latest. For him, good roads and the development of Mount Rainier National Park were bound together, for without a network of highways park visitors were tied to the railroads. At the same time, good roads transcended Rainier or any other tourist destination because they could move produce to market while they knit together the disparate sections of Washington State. When in 1907 Curtis bought a small, irrigated farm near Grandview in the Yakima Valley, he may have envisioned a paved road spanning the two hundred miles between his Seattle home and the modest acreage.

Joining the statewide goods road movement meant joining the Washington State Good Roads Association (WSGRA), the brainchild of the state's godfather of good roads, the legendary Samuel "Sam" Hill. Curtis and Hill knew one another from at least 1910 when Curtis photographed the larger-than-life Hill with a group of women at Hill's massive reinforced concrete house on Capitol Hill in Seattle. Three years later Curtis clicked a photo of a large car on one of Hill's experimental asphalt roads between Goldendale and Maryhill, Hill's floundering agricultural settlement above the barren north rim of the Columbia River. As the Maryhill flop suggested, Hill did not invest wisely in every instance, but he was a fearless financial plunger and big spender. Investments in gas and telephone companies, ranching, and the stock market were not always successful in the long run but served to support him in style for many years.[1]

More than that, Hill physically and mentally was an outsize figure, handsome, gregarious, enthusiastic, a world traveler and a linguist, a persuasive speaker and a visionary. His permanent legacy to Washington was his second enormous reinforced concrete house, begun in 1914 and now the Maryhill Museum of Art. For years it sat near his forlorn town on the windswept north bank of the Columbia, stark and brooding above its desolate surroundings, never quite finished, a home to rats, bats, birds, and bugs. Its design marked Hill as a car cultist, for long ramps on either end of the much-mocked "castle nowhere" were designed to admit automobiles up one ramp, through enormous doors to the main floor, where guests would alight in a vast reception hall, their cars to be driven through the doors at the other end and down the opposite ramp. Not surprisingly, Hill lavished time and money on the good roads movement.[2]

The origins of the good roads movement often are traced to the late nineteenth-century bicycle craze and its organized expression, the League of American Wheelmen (LAW). The LAW agitated for hard surfaced roads, but Hill cared not a rap for bicycles and the interests of bike riders. From the beginning he placed the WSGRA firmly on the side of four-wheel travel. His thinking followed the newly hatched "automobilist" agenda, personified by the dynamic Carl Fisher of the Indianapolis Motor Speedway and the patrician T. Coleman DuPont of the wealthy industrial family, who campaigned for all-weather auto roads. Like theirs, his vision ranged far beyond the few pleasure roads in the Pacific Northwest to a national network of smooth, wide highways.[3]

Whenever and however Curtis and Hill met, it's certain that Hill, almost a generation older than the photographer, befriended Curtis. Hill's expansive nature and his strong, forthright opinions, not to mention his client relationship, would have bound Curtis to him in an attitude of, if not awe, at least one of high esteem. Hill's example shaped Curtis's own advocacy of first-class highways, centralized control of road development, and a statewide view of the road problem. The older man demanded unity in a state riven by geography and conflicting interests, where the Cascade Range walled off generally rainy and commercially focused western Washington from the parched east, a larger area with a continental climate and interests centered on agriculture. Farmers often were hostile, not to adequate roads, but to highways built for what they perceived were urban pastimes like touring.

The WSGRA came along too late to reconcile these sectional and rural-urban dynamics, if indeed it ever could have done so. Hill's pioneering effort in Washington was late; it followed the first state good roads association by eight years and the first national good roads convention by five. In fact the WSGRA could preach unity and it did establish Hill's principles, but it could not impose complete agreement on the WSGRA membership, composed of road enthusiasts across the state. Outside the organization, other champions of roads agreed on the need for them but disagreed on everything else about roads, including their location, quality, funding, and organizational control.

If Hill recognized any of the difficulties, he acted as though his persuasive powers would overcome them. In a move as legendary in the good roads movement as he was himself, Hill called the first meeting of what would become the WSGRA for September 1899 in Spokane. Curtis, newly returned from Alaska, still without a reputation for his photography or anything else, would not have been on the list of Hill's prospects. Some one hundred men pledged to attend, but when Hill arrived at the meeting he discovered that only thirteen others had bothered to show up. These men and Hill became the storied "Old Guard" who persevered through years of indifference, hostility, and at least one well-organized effort to undermine their work.[4]

The great principle of the WSGRA was centralized state control of all road planning and design. As one WSGRA document expressed it, "counties, cities, and districts…should subordinate all of their plans and activities to the centralized control and comply generally with the state authority."[5] County officials, for all practical purposes in charge of road building at the time of the WSGRA's founding, resisted the association's commitment to centralization in a spectacular power play designed to wrest control of the WSGRA from the Hill group. They made their move at the 1910 convention in Walla Walla, then a farming community nestled in a rural area of remote southeastern Washington. Their plan proposed decentralized road planning and construction in several districts, effectively gutting the existing highway board.

The county commissioners may not have been quite as demonic as their actions appeared to the centralizers in the good roads association, because the issue of local versus remote control over highway building was a perennially serious matter. Farmers and farming communities wanted roads built to serve their immediate needs, however imperfectly designed they might appear to distant, practically unreachable experts. The county commissioners were right there and reachable, and if unresponsive to their constituents road concerns, could be turned out at the next election. For now, however, the focus will be on the effect of the commissioner's revolt on the WSGRA.

In 1910, the year of the revolt, anyone who paid a dollar and was a Washington citizen could become a member of the WSGRA, even up to and including the time of its annual meeting. This fact was the fulcrum of the county commissioners' coup. Three of them "drove out in their flivvers [then a term for small, cheap cars] into the country and secured forty-four members from among the farmers who were against all new road building [and] brought them in when the fight commenced." Either the farmers did not have to pay for their memberships, or their payments were refunded, "in addition [to] some inducement to come in, stay for the session and vote for the measure." During the acrimonious five-hour debate the chairman struggled to prevent fist fights on the floor. At the end, the district measure carried by eleven votes. The county commissioners won the battle but lost the war when their proposal, submitted to the 1911 legislature in the name of the WSGRA, never made it out of committee. The association tightened its membership requirements, and repudiated the district plan at its 1911 convention.[6]

The county commissioners had another reason to be worried about their future because the legislature was steadily chipping away at their traditional control of road building. The first legislation, in 1905, created a highway board and a highway commissioner, but the commissioner could only approve or disapprove of plans already drawn up by the counties. Sending the county commissioners back to their drawing boards to correct a flawed scheme was a difficult decision to make. Worse, some county commissioners ignored or evaded the law.[7]

In 1913 the legislature, exasperated with further nibbling around the edges of county authority, approved a "comprehensive" measure centralizing all highway planning in a state highway commission. Claude Ramsay, a leading light in the WSGRA and a member of the 1913 legislature, introduced the bill and shepherded it along, at last firmly lodging the principle of centralized state highway development in legislation. The county commissioners survived as road builders, though with diminished authority. Despite centralization and the loss of their design control, a hodgepodge of levies, county boards, and state allocations allowed the commissioners to continued building hundreds of miles of roads, much to Curtis's dismay.[8]

In the midst of these legislative and bureaucratic developments, the WSGRA's administrative principles served it well, maintaining its integrity despite its sometimes controversial positions and often intensive lobbying. The WSGRA dispensed with salaries, relying instead on volunteer labor. It avoided elaborate bylaws, spending as little effort as possible on rules and procedures. It mandated low dues. It declined reimbursing its members for lobbying expenses beyond the cost of personal travel and subsistence, provided that the treasury could stand the expense. It rejected large gifts from individuals or businesses, while admitting to membership any nonprofit organization interested in the good roads movement, including road building and tourist promotion groups.[9]

An original prohibition against voting membership for any "political office holder of any type" was relaxed as early as 1913, when WSGRA member and state legislator Ramsay led in strengthening Washington's highway laws. Other elected officials were members, including E. L. Farnsworth, the 1916 president. The annual meetings decided policy in the form of resolutions, typically favoring the expansion, physical enhancement, and improved funding of the highway network. Between annual meetings authority rested most conspicuously with the president, a situation dovetailing with Curtis's penchant for assuming control in ambiguous situations. For advice and assistance the president could call upon the vice presidents, one for each Congressional district, and the executive committee, also district based. All were elected by convention ballot upon recommendation of the nominating committee.

Hill's outsize personality dominated the association in its early years. He presided over the WSGRA from 1899 to the 1910 annual meeting. His first and enduring emphasis was on improving the quality of rural life by making markets accessible through hard-surfaced roads. If farmers could spend significantly less on transporting their produce over good roads, not ribbons of muddy ruts, they could keep a greater proportion of their income. They would use their increasing returns to make rural "life attractive and comfortable," especially for the rising generation of "ambitious" youngsters. Then "our girls" would "stay on the farm and become the mothers of a virile race of men," not head for the city to be "manicurists" and "stenographers," or worse, "variety actresses." Boys who left the farm would "succumb to the lure of the Great White Way or become chauffeurs or clerks."[10]

The farmer in Curtis agreed with Hill's concern for quality roads in rural areas, although his sense of statewide unity also embraced urban prosperity. The people required to staff burgeoning city businesses had to come from somewhere, thus Curtis raised no argument against rural youth moving into what Hill believed were the moral cesspools of the cities. Rural prosperity would induce enough young people to stay down on the farm. Besides, Hill was no beacon of morality, as his thinly concealed mistresses and illegitimate children revealed. His anguish over the compromised purity of rural youth could have seemed to Curtis as nothing more than a hypocritical concern for saving lesser mortals than himself from temptation.

Hill's comprehensive view of the road problem also influenced Curtis. It was a vision with many facets. Automobile touring never displaced rural improvement in Hill's mind but it did become an equally important justification for good roads. As Washington's roads improved, however marginally, and cars became more reliable, Hill took in great gulps of scenery from his powerful Locomobile. He was not, however, a back-to-nature enthusiast who saw car travel and rustic camping as a brave revolt against the rigidities of trains and staid hotels. His thinking and activities encouraged Curtis, whose enthusiasm for touring initially focused on accessibility to and within the Rainier national park. Despite Hill's interest in scenery as opposed to nature, his viewpoint appealed no less to the back-to-nature buffs, who wanted roads to smooth their paths to the pure.[11]

Farmers rejected, not roads, but Hill's tourism and the back-to-nature boom that came in its wake. Cars were noisy, disturbing rural tranquility and farm animals. Arrogant drivers sped, their cars endangering life, menacing property, and throwing up clouds of choking dust. Cars' drive wheels gouged deep ruts into roads surfaced with original style macadam, tiny rock fragments held in place by frequent sprinkling. Their automotive vandalism contrasted with the benign, compressing action of wagon wheels. Auto tourists and the back-to-nature crowd showed scant interest in respecting rural property, leaving behind picked-over crops, cans, bottles, campfires, blowing papers, and evidence of their careless sanitation.[12]

Seen in the light of rural disgust with urbanites, the failed revolt of the county commissioners against the WSGRA's centralizing thrust was not really a lining up of farmers "who were against all new road building." There is no evidence for a farmers' rebellion in 1910. In other words, farmers may have been opposed to pleasure highways for urbanites but not against rural roads planned by rural people for rural purposes. There are hints, many years after, from veterans of the good roads crusade that they struggled to win over galluses-snapping rubes, but that is all the evidence for a rural rebellion against roads. It is possible that the greatest effect of the 1910 commissioner's insurgency was a role in convincing Hill to leave his presidency and his close ties to the WSGRA, and to diminish his interest in farm roads in favor of long distance highways. By 1912 his attention had shifted to shaping what became the stunning Columbia River Highway on the south, or Oregon side, of the magnificent watercourse.[13]

A falling out with Governor Marion E. Hay may also have influenced Hill's abandonment of sustained interest in Washington. The political donnybrook perplexed and disturbed Curtis because it threatened the primacy of first-class highways in favor of a greater mileage of indifferently produced roads. Hill precipitated the break in May 1911 when he wrote Hay a scorching letter denouncing the governor for his road policies, an attack resulting in Hill's leaving the chairmanship of a powerful advisory board on roads. Hay replied, defending himself as a strong good roads advocate while declaring Hill's methods to be wasteful and unsuited to conditions. That was not all. The governor suggested that the commissioners' revolt at the 1910 convention was more than a power play, strongly hinting that their tactic went deeper than simply bribing "forty-one farmers who cared nothing about roads." No ignorant, contemptible hayseeds, the farmers were canny people who understood the limits of Hill's plan to build expensive roads and build them all at once. "Even you yourself," Hay declared, "with a number of personal friends, were unable to stem the tide, and went down to defeat at the last State Good Roads Convention," while only firm control of the 1911 meeting "saved the day" for the centralizers.[14]

Worse for Hill, Hay decided to do away with convict labor on state roads despite what seemed to be his public promise to continue it. Hill argued that convict labor was efficient (convicts who weren't efficient would be sent back to prison), that the convict's wages were paid to them or to their families, and that convict labor was safe, despite fears of some miscreants breaking loose and running amok through the countryside. On the other hand, opponents declared, convict labor deprived regular road employees of their rightfully higher wages and qualified contractors of their profits. Experienced employees were superior, or would become superior, to convicts. There were hints of abuses, such as working convicts for "starvation" wages.[15]

Hay's decision to abolish convict labor was the last straw for Hill. In 1912 he opposed Republican Hay and backed Democrat Ernest Lister with all the influence and rhetorical skill at his command. "We can't make bricks without straw, but we can make roads without Hay," he allegedly declared. Hay lost and Hill was not shy about taking a lot of credit for his defeat, but it may have happened for other reasons. Republican Hay was running in a year when Theodore Roosevelt's Progressive Party or "Bull Moose" campaign split the Republicans. A historian of the Pacific Northwest noted Hay's ambivalence about the then-popular progressive reform, as well as his "drab personality" and lack of the "political flair" common to reformist governors. Besides, Hay never was elected to be governor, but as the lieutenant governor had served most of a full term after the governor died.[16]

Curtis was upset by the antagonism between Hay, his party's leader in the state, and Hill, the preeminent promoter of good roads. During the campaign Hay visited Seattle, where Curtis spoke with him about the discord between the two. After returning to Olympia, the state capital, Hay sent Curtis a copy of his 1911 missive to Hill, along with a covering letter. Hay wrote to Curtis of Hill's "great work in arousing the people of the state to the necessity and economy of good roads. However... he is far from practical." Hill favored unpopular highway funding and "extravagant and wasteful methods" of road construction. Leaving him in a position of influence "would have meant the ruin of the good roads movement in this state." Curtis understood the difference between building excellent roads and the effective administration of a comprehensive road program. "I wish to thank you for putting me right in the matter of roads and particularly the employment of convicts in the construction of state roads," Curtis responded.[17]

Hill later returned to the good roads fray in Washington, this time adding his erstwhile acolyte, Governor Lister, to his rogue's gallery of good roads apostates. It mattered little that Lister tried to find enough prisoners for road work or proclaimed Good Roads days. He failed to use convict labor and cancelled work on a north bank Columbia River highway, one of Hill's pet projects. In 1916 Hill launched a speaking tour around Washington, a campaign that served to highlight profound differences over road construction. While continuing to lambast Hay, he also condemned the chairman of the state senate road committee and what he believed were the misdirected policies of the Lister administration, which emphasized inexpensive rural roads instead of high quality highways. "After the name of Lister, because of these roads, must be written at the close of his administration, Failure, Failure, Failure." Hill's attack prompted a response from E. L. Farnsworth, then the president of the WSGRA. "Mr. Hill is rich and likes things nice," Farnsworth wrote, "but the people of Washington and their legislators opted for an extensive mileage of serviceable roads, not a few miles of first-class thoroughfares such as the magnificent Columbia River Highway in Oregon." Hill was properly proud of the Oregon highway but it cost about $48,000 per mile instead of the under $12,400 per mile for Washington's difficult Snoqualmie Pass road.[18]

Lister answered Hill, too. Farmers and "city people," he wrote, "while appreciating good highways, much prefer the several hundred miles of roads [built during his administration] to sixty or seventy miles of automobile boulevards." Lister continued: "However, it is impossible for anyone to control Mr. Hill's expressions, whether right or wrong, when he makes up his mind to state them." Indeed Hill's censure of Lister may not have been so wide of the mark. F. J. Wilmer, in a reminiscent mood when he stood before the WSGRA to deliver his 1933 presidential address, was not kind to Lister's pioneer efforts. Wilmer credited Lister with the "first real road building" in the state, but criticized his concern for mileage at the expense of careful engineering. "The poor alignments that afterward required wholesale revisions, and the steep walls that made spring slides for years, were results of Governor Lister's ambition to build miles, and yet more miles." Wilmer was right, probably, but he overlooked what Lister's constituents wanted—miles and more miles. Hill's outbursts nevertheless distanced him from the WSGRA leadership and important people in the state government while undermining his personal popularity. He remained in the movement, a Grand Old Man to be rolled out like a tin god on critical or nostalgic occasions, but his influence declined.[19]

Curtis abandoned Hill's banner because the flamboyant entrepreneur failed to lead toward a constructive policy, though they remained on good terms. In 1918 Hill wrote to the popular historian at the University of Washington, Edmond S. Meany, about his strong sentiment for Curtis and others. "I hope the names of Meany, [Reginald H.] Thomson, Curtis and Hill will always be joined together as our hearts are."[20]

Curtis's star rose in the WSGRA while Hill's descended. Though physically unimpressive, Curtis was a forceful speaker and, unlike Hill, a good negotiator, compromiser, and analyst of road data. He and other younger WSGRA men benefitted from three developments: the Federal-Aid Road Act of 1916, the Federal-Aid Highway Act of 1921, and the 1921 state gasoline tax.

The Federal-Aid Road Act of 1916 was significant for establishing the principle of partial federal funding of highways on a dollar-matching basis, and advancing federal-state cooperation far beyond the limits of previous federal efforts. The act set the formula for federal aid to each state, one-third based on area, one-third on population, and one-third on the mileage of rural roads. Every state had to have a highway department to select and perform the engineering work on projects, subject to final federal approval. Once the plans were federally approved, states would be responsible for supervising construction of the highways, and for maintaining them. If any organizations effectively fought for the 1916 act, they were the American Automobile Association (AAA), and the American Association of State Highway Officials (AASHO), not the WSGRA or any other good roads group.[21]

Though it broke new ground and many of its provisions would survive in later legislation, the 1916 act had little practical effect. Every state but California had to modify or, in a few instances, to create highway commissions that met federal standards, causing delay. Conflicts between federal and state engineers held up many projects. The act emphasized mail carrying routes, roads that failed to forge a connected, interstate system, as some backers of the bill hoped. Worst of all for the 1916 measure, the country's entering World War I dropped road building to a low priority. When shippers turned to trucks to supplement snarled railroads, the trucks' brutal, hard rubber tires quickly damaged or destroyed existing roads. Road users unfairly blamed the federal bureaucracy for the highway failures.[22]

The Federal-Aid Highway Act of 1921 repeated many of the 1916 provisions, including the aid formula, ultimate federal control, and state responsibilities. Curtis and his allies in the WSGRA had little to do with passing the 1921 act, any more than they did the 1916 legislation. They welcomed it, although Curtis paid practically no attention, then or later, to the details of federal highway acts. Federal activity interested him only insofar as it provided funding or strengthened centralization in the state highway department. The 1921 act did both. It gave state highway commissions control over all roads designated farm-to-market, reinforcing Washington's highway laws. Under the new federal provisions Congress increased matching funds from 1916's $75 million over five years to about $75 million every year. If Washington were to share in the federal largess, it had to increase its road building appropriations, which it did. In the biennium ending in 1924, the state spent over $14 million while the 1930 biennial expenditure topped $23 million. Those figures included federal aid as well as income from a new source, a state tax on gasoline.[23]

The WSGRA supported the 1921 tax but appears to have had less to do with its passage than the example of adjacent Oregon, where the legislature and governor approved it in 1919. Unlike some other taxes, the gas tax proved relatively popular, easy to initiate, collect, and increase. It seemed fair on its face, because it was a "user" tax, with those who traveled the highways paying their reasonable share of road maintenance and improvement. It appealed to the belief in progressive taxation; drivers of "gas guzzlers" (to borrow a later phrase) who could afford to operate them bought more gasoline and paid more taxes than did owners of more modest cars, miles traveled being equal. The tax was progressive in another sense, for those who roamed the roads more frequently paid more than those who drove over them less. Nevertheless, some realities of the gasoline impost, as they existed in Washington in Curtis's era, modified notions of its uniformity and fairness (see Chapter Five). However viewed, the gas tax was only a part of the money available for highways. The 1923 legislature established a motor vehicle fund, its components varying over time.[24]

The only constant in the motor vehicle fund was the rising gas tax, initially one cent per gallon on distributor sales. In 1923 the rate rose to two cents, and three cents in 1929. The 1931 legislature increased the tax to five cents, an especially onerous exaction during the Great Depression. The tax remained at five cents for the rest of Curtis's life. The gas tax increase did not mean that the state spent four times the dollars on major roads in 1931 and later as it did in 1919.[25] The complexities of population and highway mileage increases, changing price levels, and improving road surfaces, not to mention the bewildering variety of alterations in the motor vehicle fund and the equally confusing pattern of its disbursements, made it almost impossible to know which shell the tax burden or tax benefit pea was under. This shouldn't suggest that the state legislatures, state governors, the WSGRA, or anyone else was playing a deliberate shell game with taxpayer funds. All were responding to the success or failure of previous efforts, the promise of new initiatives, pressures from the highway commission or from lobbying groups such as the WSGRA itself.

In any case the new tax seemed, at the outset at least, to be nothing but a boon. From the perspective of 1921, federal legislation and the gasoline tax capped the success of the good roads movement and its startling progress in two decades. The WSGRA maintained its integrity and independence while seeing its principle of centralizing all important road building decisions come close to complete realization. Though the WSGRA could take little direct credit for the federal acts or the gas tax, its work was undeniably part of the larger victorious struggle. Curtis could believe that, not only was centralization achieved, but that the ground was prepared for a broad, unified development of the state's highway program, and for concentration on first-class paved highways. His hindsight from the vantage of 1921 would see almost as much progress as an imperfect world would allow. Foresight from the same vantage point would predict a close-to-cloudless future for the WSGRA. In fact, the association was entering two difficult decades, trying years best seen through Curtis's view of events.

Chapter Four

The Good Roads Movement: Battling Roland Hartley

Frustration, conflict, and anger sometimes marked Curtis's struggle for improved highways, but never more than when Roland H. Hartley occupied the governor's office, 1925 to 1933. Hartley could be intolerant, intransigent, and vituperative, calling his critics and political opponents "pusillanimous blatherskites," and "goggle-eyed jackasses." Most galling to Curtis was Hartley's insistence on drastic economy in state government, including highway expenditures. Hartley was in office for less than a year when Curtis wrote that "there is a great deal of foolishness connected with the majority of the governor's ideas."[1]

Had Hartley been merely a contentious, cheeseparing politician, he probably would never have gained, or at least not retained, the governorship. He was, however, highly intelligent, shrewd, and flexible when political realities dictated flexibility. In 1927, when his parsimony and obstreperousness threatened his governorship, he became "utterly charming," and responded with a substantially increased budget, including more money for highways. He could be congenial and gregarious, and was almost always willing to forgive and forget following a dispute, no matter how acrimonious. In 1932, during the Great Depression when government parsimony was no longer in vogue, Hartley lost a bitterly contested fight for the Republican gubernatorial nomination. When next he saw his victorious rival, he smiled, shook hands, and engaged him in friendly conversation.[2]

Whatever his specific actions, Hartley believed corporate management structure to be the ideal system for any large organization, public or private. For him, placing state government on a business basis did not mean simply controlling costs, it meant converting the state into a corporation with the governor as chief executive, insofar as it was possible to do so. The state was not a corporation, of course. The legislature could be unruly, the courts were independent of the governor's control, and the voters sometimes got things wrong. In some instances it might be necessary to cooperate with or appease others, but such concessions were tactical only. The voters should support the governor as they would if he were a corporate head, and anyone who got out of line should be brought to heel as he would be in any responsible corporation. Curtis's criticism of Hartley, that he wanted "the entire highway fund turned over to him to be expended whenever and wherever he pleases," or that "many communities are afraid...he will cut off appropriations for their roads," would have been pointless to the governor. These were results to be praised; this was how any reasonable executive quieted opposition within his organization.[3]

Prior to the Great Depression, Hartley was popular with the people of Washington. Having been mayor of Everett, a lumber milling town north of Seattle, and a member of the state legislature, he announced for the Republican nomination in 1916, and again in 1920. He did well, but not well enough until 1924, when he eked out a close win over the Republican organization's candidate, then crushed the Democratic nominee by almost 96,000 votes, a huge margin at the time.[4]

Hartley won because voters generally approved his program of tax reduction, simplification and centralization of state government functions, and his fearless attacks on prominent state institutions, especially the University of Washington. They liked his feisty platform manner, aggressive campaign style, and the not always printable epithets sprinkling his speeches. Hartley was fiercely loyal to his friends and inspired loyalty in return, even when he shifted his stance to meet political exigencies. Friends stuck by the man himself, while opponents were obliged to hit a moving target if they could. In 1928 Hartley won the majority vote in the Republican primary, and bested his Democratic opponent after a colorful campaign invoking—for the first time—extensive radio politicking.[5]

On the face of things there were reasons why Curtis and Hartley should have been political allies. Both were staunch Republicans, short of stature, physically active, and admirers of Theodore Roosevelt. Both knew Minnesota, Hartley moving there some fourteen years after he was born in Canada in 1864, while Curtis was a Minnesotan by birth.

Both knew hardscrabble times, raised in families whose farmer-preacher heads could not make an adequate living either farming or preaching. Neither enjoyed much formal education. Curtis's classroom days ended with primary school, while Hartley gained the equivalent in a hit-and-miss manner. There were differences in their career trajectories, occupations, ages, attitudes, and social preferences. The mature Curtis lived comfortably enough, but Hartley was "at least a millionaire" by the time he became governor, making him the owner of a substantial fortune in 1925. None of these personal distinctions should have prevented their close cooperation.[6]

More than that, Curtis would have warmed to Hartley's campaign rhetoric. Hartley, a member of the close-knit lumber management group in Everett, articulated a profound contempt for labor unions. While not so alienated from unions as Hartley, Curtis believed in such virtues as personal independence, voluntary cooperation, and hard work as the solvents of social problems. Government, Curtis understood, had important obligations but Hartley's calls for governmental efficiency would have resonated with him. He would have warmed to Hartley's 1924 platform, which declared "good roads" to be "vitally essential," and endorsed the "existing road building program." Current construction would result in "the finest roads in America. That program must be carried to completion." There is no record of how Curtis voted in the Republican primaries, but he should have been satisfied with Hartley's victory in 1924.[7]

Curtis's presumed contentment with Hartley quickly evaporated. In his inaugural address of January 1925, the new governor called for elimination of the levy on all property for county highway construction and maintenance, and shifting a similar sum from the motor vehicle fund to the counties. Highway users, he declared, should pay for highway programs. The legislators quickly obliged him, despite some opposition. They passed a few other measures at his bidding. Then, as Hartley requested, they adjourned to await the results of a comprehensive "business survey" he planned for all state institutions. The solons again obliged, reconvening in November at Hartley's call. His address to the legislature was, to put it mildly, controversial. Among other far-reaching measures, he favored reducing the motor vehicle fund, the source of highway money, because, he declared, there was more than enough money available for highways, and no need to increase the fund's major component, the state gas tax of two cents per gallon.[8]

Hartley doubted that any great public demand for highways existed, because if the people knew how much was spent on Washington roads, they would be horrified. If they really wanted "hard-surfaced joy roads," then the state should "reduce the pot and leave the people's money with them for their own use and expenditure." He did not explain how the increased private funds would provide satisfactory public roads, but in any case there was no "economic necessity" for wild spending on highways. The public demand was a fake, a phantom of the highway lobby, "the cement crowd, the material men, the machinery folk, the contractors, the automobile club secretaries, and the great army who are living off of, some of them growing wealthy from, highway construction."[9]

Hartley's assault on highways was critical for Curtis and many of his associates in the WSGRA. Some of them came to detest Hartley, but they shared with him an important agreement in principal. Both the governor and his critics worked to centralize control over highway expenses in Olympia, though they fell out over who would do the controlling, how it would be done, and who would benefit. For example, the governor saw no harm in appropriating a large sum to the state's county commissioners to spend for road improvement.[10] Such largesse, combined with the unspoken but obvious threat of the line item veto, would bind the county commissioners to the governor.

Curtis opposed funneling any substantial sum from the state motor vehicle fund to the counties for local roads. When his solution proved to be politically impossible, he argued against increasing allocations to the counties (or cities) because every dollar granted to them meant one dollar less for a centrally planned system of primary highways. If a county wanted strictly local roads, it could levy the necessary funds against property in the county, or it could pass a bond issue. Too much county money was "spent on a hit or miss plan...with a resulting loss that is staggering." The solution was centralization, with each county "required to put their County system on a definite plan for county cooperation with the State."[11]

Hartley and Curtis first crossed swords over Hartley's decision to dismiss James Allen, an effective, "highly popular" highway engineer. Curtis, writing as chairman of the good roads committee of the Seattle

Chamber of Commerce, urged Hartley to retain Allen, enclosing a resolution of the chamber's board of trustees to that effect. The road committee, Curtis wrote, had found Allen, "in his management of the State Highway Department…as free from politics as it is possible for any state official to be." Hartley was unreconciled to political independence from any state employee, and especially not from a gubernatorial appointee as powerful as the highway engineer. He accepted Allen's resignation, replacing him with J. Webster Hoover, an "unknown" from Hartley's adopted home town of Everett.[12]

By 1926 Curtis was smarting over the effects of Hartley's "obnoxious" false economy. An undermanned, underequipped highway patrol couldn't prevent overloaded trucks from destroying Washington's roads. There would be fewer accidents, Curtis asserted, if an effective highway patrol enforced safety regulations. The negative impact on highways of repealing the levy was equally unfortunate. The result, combined with the over-funding of county road building, was inattention to the state's main highways. The Sunset Highway, the major direct link for the eastern and western sections of the state, was in "bad condition," while the north-south Pacific Highway required paving "to relieve congestion of the other roads." In short, "the State Highway system is in the most grievous danger it ever has been since the inauguration of the real system in 1913."[13]

Curtis worked hard to get two short roads added to the state network even though he and other highway boosters were opposed to the indiscriminate expansion of an already overwhelmed highway system. For him, the more important was an eleven-mile extension of an existing state highway to connect with a short stretch of road through the nearby national forest and finally to a Mount Rainer park road. The road then would run to the projected West Side Road, essential to open the western area of the park to automobile traffic. The eleven miles were also a necessary link in a hoped-for circumferential road around the park. At a meeting with the district representative of the Bureau of Public Roads (BPR, in charge of roads in the park), and the district forester, the state highway commission agreed to the extension. Hartley attended, joining the other two commission members in a unanimous decision in its favor. Curtis was exultant. "I think this settles this very complicated situation and that the last political difficulty related to the West Side Road is out of the way," he wrote.[14] He would be sorely disappointed when the legislature later extended a state highway to Rainier park, but Hartley vetoed the appropriation, ignoring his earlier approval of the road (see Chapter Seven).

In 1927 Hartley was ready to move boldly on the highway front, approving increased expenditures to appease some critics, while denying funds to others. He did so under the rubric of a unified biennial budget. The new budget system resulted from the legislature's enthusiasm for most of the reforms Hartley suggested early in 1925, well before he unleashed his economizing and centralizing proposals. The 1925 law provided for a comprehensive budget drawn up in the state department of efficiency, then presented to the governor for his approval and finally to the legislature for its consideration. Previously, each department prepared its own budget for submission to the legislature. The unified budget was not designed to end the give-and-take of the budget bill process or to prevent department heads from appearing before legislative committees on behalf of their departments, but rather to present the legislature and the governor with a complete, comprehensible, budget picture. Hartley interpreted the unified budget as coming from the governor to the legislature on a take-it-or-leave-it basis.[15]

Hartley's concern for centralizing the budget may have been the same as his other moves toward strengthening the executive branch—to rationalize, simplify, and economize. Whatever his initial understanding, at some point he perceived a political application, the line-item veto, to quell opposition by denying road money to counties "where Hartley support was weakest."[16] Hartley did not confine himself to cutting the legislature's highway appropriations for disaffected counties. Indeed he "took a healthy swing at Kitsap County, striking from the highway budget every appropriation for road construction in that section," the *Seattle Times* declared. Kitsap was pro-Hartley country but the governor's motive in this instance was the dislike of the county's state senator, the newspaper asserted. "Funds for road building, in the opinion of the governor, are rewards for good conduct. Naughty anti-Hartley counties get nothing!" On the other hand, Hartley vetoed three bills sponsored by political allies while approving critical highway money for Seattle and King County, although neither the city nor the county areas outside Seattle were consistently strong Hartley bastions. Though the operations of Hartley's mind were not always transparent, it was obvious that he lived by another calculus than simply rewarding friends and punishing enemies.[17]

For Curtis, Hartley's institutional battles loomed larger than any struggle over a highway. They began in 1927 when Hartley dismissed his own appointee, Hoover, who had shown an unsuspected streak of professional independence, in favor of Samuel J. Humes, a Seattle engineer. Humes soon requested the resignations of the top Hoover engineers. These actions infuriated the other members of the highway committee, State Auditor C. W. Clausen and State Treasurer William G. Potts, both long-serving elected officials. Clausen and Potts were no friends of Hartley's because he had excluded them from their accustomed advisory roles and had criticized them personally.[18]

By tradition the governor served as chairman and the highway engineer as secretary and unofficial consultant to the three-man committee. Clausen and Potts challenged this arrangement. Clausen nominated George T. McCoy, an engineer who Humes had just fired, to be secretary, whereupon Hartley announced that Humes would be secretary. "McCoy is out of it. He's through," the governor shouted. "I am chairman of this committee, and I am going to be chairman in reality as well as name." Clausen enlivened the meeting when he called Hartley a "damned liar."[19] When Potts and Clausen voted Clausen as chairman of the committee, Hartley ruled the motions out of order and the meeting broke up. Hartley then went to the state administrative committee, where his appointees named him chairman and Humes secretary. The attorney general soon overturned the appointments, ruling first, that the administrative committee had no authority to name the officers of the highway committee and second, that the committee members could name their own chairman and secretary. Clausen carried his case against Hartley to the state supreme court, while McCoy asked the same court to rule against Humes, who had requested the committee's records on Hartley's orders. Meanwhile, Clausen agreed to allow Hartley to preside so that highway construction bids could be opened and awarded. At the end of a second meeting to award more bids, another disagreement ended when Humes physically ejected McCoy from the room.[20]

The WSGRA stepped in during the donnybrook. On May 12, a committee of eight chaired by the formidable Samuel Hill but not including Curtis, requested "that your differences be immediately settled." Hartley agreed, so long as the committee's modus vivendi survived. Clausen and Potts did not agree, and boycotted several meetings. The committee continued its fractious ways until a supreme court decision of June 17 which held, in effect, that the group could select its own chairman by majority vote and that its minutes could be kept by anyone whom the majority chose. Hartley conceded the chairmanship and the secretaryship but continued to be obstreperous. When Humes defied a supreme court ruling to return the committee's records, he was jailed, briefly, to Hartley's boundless anger.[21]

Curtis had had enough of Hartley even before the supreme court's ruling. He correctly predicted that "Hartley will be renominated" in the Republican primary unless Republicans united behind one man "strong enough to beat Hartley." Hartley garnered more primary votes than his two opponents together. "Personally," Curtis had declared, "if Hartley is nominated I will vote the Democrat ticket." Curtis may have broken with his party and voted for Hartley's opponent in the general election, the flamboyant A. Scott Bullitt, but his vote did not matter.[22]

Hartley's reelection did nothing to calm his feisty behavior, but it did cloak him with the authority to ask the legislature for an end to the disputatious highway committee. In 1929 he proposed a highway department headed by a director of highways, appointed by the governor. The legislature concurred, and Hartley named Humes to the post. Hartley's and the legislature's actions prompted the WSGRA to authorize appointing a committee to study the highway laws of Washington and other states, and report its recommendations to the 1930 convention. The WSGRA named a nine-man committee, Curtis included, without advance consultation and without providing any expenses. All agreed to serve. The committee reported its labors incomplete in 1930 and requested another year. It also wished to avoid presenting its recommendations to the 1931 legislature in favor of an initiative referred to the voters in the 1932 election. The preference for an initiative was sensible, given the likelihood of a Hartley veto of the sweeping changes proposed, even assuming the legislature's unlikely acceptance of a reduction of the governor's power over highways.[23]

For the report was radically anti-Hartley. It proposed scrapping the existing director of highways appointed by the governor (Humes, appointed by Hartley) in favor of a paid five-member commission representing separate districts, and appointed by the governor, plus six elected state department heads. The proposal would have undermined the governor's power by giving the commission the authority to appoint the highway engineer, and by placing the state patrol under the commission. It required the legislature to approve a "lump sum…in lieu of…appropriation under the present budget system," ending the line-item breakdown of previous highway bills. This proposal would force the governor to accept the whole road budget or run the high political risk of vetoing the entire package. The accompanying report was mostly restrained and judicious, comparing the Washington highway structure with those in other states, and carefully defending its preferences.[24]

The report's language left no doubt of Curtis's strong hand in its authorship. Such phrases as "one man control," "centralizing authority in the chief executive," "arbitrary control," and "control the politics of the state with the taxpayer's money," reflected Curtis's major concerns. Their inclusion was not surprising, given Curtis's conviction that Hartley was "unbalanced," or his belief in Hartley's "definite hostility which reaches even into business transactions" (presumably fewer or no Curtis photographs of Hartley-related mills in Everett).[25] The report, after mentioning three line items Hartley vetoed in 1927, went on to relate how "the day before the Governor vetoed these three items amounting to $62,000.00, because they were not in his budget, he approved items in the highway appropriation bill amounting to more than $2,300,000.00, none of which appeared in his budget." The last statement resembled sentences in Curtis's private letters concerning Hartley's fiscal practices.[26]

The recommendations exploded a bombshell in the WSGRA's 1931 convention. A vote to accept the committee's report passed 237 to 202 on a written ballot, a narrow margin in view of Curtis's high profile in the organization and the elaborate statistics, argument, and protestations of non-partisanship accompanying the recommendations. On February 12, 1932, the WSGRA's executive committee dealt the proposal a near-fatal blow when it refused to commit the organization to participating in any initiative ballot measure including the proposals. The vote was twelve to nine, decisive considering that four members of the study committee, including Curtis, also served on the executive committee. No account of the decision survives in the printed record but some inferences may be drawn.

First, the WSGRA always maintained that its interests were in good roads and not in political disputes. The proposals sailed the WSGRA close to the thin ice of partisanship. A disgusted John Carlson, a county commissioner and delegate from Kitsap County, told his constituents that the 1931 conference "developed into a political meeting for and against Hartley," and that the WSGRA "was of no further use as it had now deteriorated into a political convention." Any initiative signature-gathering and campaigning was bound to be expensive and divisive, projecting the WSGRA into the rough-and-tumble partisan arena.[27]

Second, by the fall of 1931 Hartley was in full retreat from his anti-highway position. He approved a state road to Rainier, though too late for his concession to make any difference to park road planning. He spoke to the 1931 good roads convention, not something he always did. His director of highways, Humes, turned out to be an effective administrator. In 1929 the WSGRA had expressed its confidence "in Honorable Samuel J. Humes and his efficient corps of engineers." The first resolution of the organization's 1930 convention praised "the splendid work of the Director of Highways and his entire corps" for progress "along sound and economic lines...accomplishing the largest per annum of work in the history of the state."[28]

Eventually Curtis climbed aboard. At first, in 1927, he complained of Humes' maintenance program as being "little short of criminal." His anger seemed to be directed at a popular road oiling program designed to lay down a petroleum binder over rock and gravel roads, making them "dustless." Defeating dangerous clouds of choking, blinding dust thrown up by oncoming cars was a major concern for motorists. In an era when the open touring car remained popular for summer travel, buttoning on stifling side curtains was the motorist's only barrier against an unpleasant layer of dust over clothing and skin. Rolling up the windows of a closed car at a time when auto air conditioning was not even a dream was an equally unacceptable defense during eastern Washington's hot summers.

Curtis initially thought oiling was temporary and uneconomical at best, no substitute for a permanent surface. In 1930, however, he came around to supporting Humes. "It is generally rumored that there is great extravagance in the Highway Department but I feel that for 1930, we have secured good values." The next year he wrote a long letter to Mark E. Reed, a prominent lumberman and retired legislator who had fought his own battles with Hartley. Reed had accused the Humes highway department of rebuilding some highways unnecessarily and uselessly relocating others. Reed was right in some cases, Curtis wrote, but in others "you are doing the Highway Department of the State an injustice." Humes and his force had to reconstruct highways damaged by trucks and realign others poorly located.[29]

The third possible reason for the executive committee's refusal to involve the WSGRA in an initiative ballot measure lies in the political-economic realities of the Great Depression. By the time of the executive committee's February 1932 decision, the economic outlook was bleak. The future of the Republican Party, saddled with the responsibility for the depression disaster, was grim. Hartley developed no policy response to the depression, except to affirm his belief in removing government from the economic scene as much as possible, an unpopular attitude as his defeat in the Republican primary affirmed. Politically Hartley was a dead horse; better to stop beating him with a proposal to sap his waning strength, and to await events. Democrat Clarence D. Martin swept into the governor's office in 1933 and vindicated the executive committee's decision when he appointed Lacey V. Murrow to be director of highways. It was an inspired appointment. Despite Curtis's lingering hopes for a highway commission, in 1933 President F. J. Wilmer effectively took the WSGRA out of the controversy when he declared for "one man authority and one man responsibility" in road administration.[30]

The battles with Hartley marked Curtis's rise in the WSGRA. He ascended because of his willingness to put in long hours of hard work for no recompense beyond the possible triumph of his views and the approbation of his fellow laborers in the good roads vineyard. His fluency in speech and writing, his gift for analysis, and his candor, combined with an ability to compromise on most issues, propelled him to important posts in the WSGRA. His election to the presidency in 1933 was not, for all that, a sure thing. The fractious debates over his effort to enroll the WSGRA behind a political assault on Hartley resonated in the nominating committee, which required four ballots to nominate him, and then only by 16½ votes to 14½.[31] Luckily for Curtis the convention rarely if ever disputed the committee's selections. Like other presidents he accomplished most of his work in the following year and gave his presidential address in 1934.

Curtis's work involved highway problems beyond confrontations with his nemesis Harley. By 1930 he had risen to similar positions in the Seattle Chamber of Commerce (see Chapter Seven). These moves, combined with his chairmanship of the Rainier Park Advisory Board, allowed him to coordinate the organizations' responses to a variety of issues. To cite two instances, the WSGRA, the advisory board, and the chamber pressured King and Pierce Counties, state legislators, and governors to improve connecting roads to Rainier. The WSGRA and the chamber cooperated to publicize an "international highway" from the Canadian border north through British Columbia and the Yukon to Fairbanks, Alaska.[32]

Hartley and organizational convergence aside, Curtis resolved the highway problem into questions of strategy and tactics. He did not use the military terms but he saw the strategic problem to be the state's failure to pave its primary highways with concrete. This strategic failure hurt tourism. When auto tourists from Yellowstone National Park and other points west reached the Walla Walla area in far southeastern Washington, they turned southward to Hill's and Oregon's paved, sumptuously scenic Columbia River Highway. Given "a paved road leading south to the much advertised Columbia Highway and a dusty road leading to Puget Sound, you can scarcely blame the tourists for continuing on the pavement," he wrote. Paving a northwesterly highway from Walla Walla would be "of tremendous importance to the [Rainier] National Park." As it was, on a recent visit to Yellowstone "I could not really urge those I met…to make the trip via the northern route."[33]

The damage to tourism, not to mention all other highway uses, was serious enough. It could have been mitigated by an aggressive program of paving state highways, but tactical flaws prevented implementing an effective paving strategy. The tactical shortcomings included large payments to counties and cities, payments that could be used to pave state primary highways, over-reliance on the "pay-as-you-go" method of financing (see below), instead of bonding to supplement pay-as-you-go, and diversions from the motor vehicle fund to finance non-highway activities. Curtis's overriding problem was this: he could criticize poor tactics but there were too many arguments in their favor for them to be abandoned. He was in a box from which there was little chance of escape. As he repeatedly complained, the counties received a large proportion of the available highway funds.[34] The cities, according to the highway department's official history, garnered funds based on their assessed valuations, funds not tied to road maintenance or construction but which could be used "for bond retirement or other similar purposes."[35] Other obligations, including maintenance, left relatively little for new construction, only part of which involved paving. As Curtis told the 1934 WSGRA convention, "excessive maintenance" of unpaved roads could be reduced through an active paving program of one hundred miles per year. Yet the state funds paved but seven miles in 1933. Given the 2,810 unpaved miles in the primary system, "it would require 400 years to complete the job. This is a longer time than even I expect to be interested in roads."[36] Curtis's sally produced some appreciative laughter, probably, but his point was well taken. Depression-induced federal funds above the usual amounts had been flowing into Washington since 1931, but they did little to bolster the state's skimpy paving record of 1933.[37]

Curtis inveighed against the state's generosity to the counties in the post-Hartley era, because the county roads were often unnecessary and ill-planned, and because their waste and extravagance should be dropped in favor of responsible work on state roads. He told the 1933 WSGRA convention how he did "not overlook the needs of the counties and the cities," but knew that King County was "paving county roads which had less traffic than many state roads that are not paved. In other words, the county system is being advanced at the expense of the state system." His 1934 presidential address was even less complimentary to the counties. "A comparative analysis of results obtained during the last ten years shows a woeful lack of accomplishment on the part of the counties," he asserted. Effective coordination of road construction "was scarcely possible with 39 separate and independent governing bodies directing the work." Therefore the state should take back "the entire motor vehicle fund," building and maintaining primary and secondary highways through counties and cities where necessary.[38] Curtis overlooked the fact that what he termed extravagant and uncoordinated building resulted from the centralized disbursement he so desired. Perhaps his vision included refining the existing highway legislation to allow cheaper county roads but he did not say so.

The view from the counties was not Curtis's. In Kitsap County, the meeting minutes of the North End Improvement Council (NEIC) reflected a different reality. Kitsap, then overwhelmingly rural with the exception of Bremerton, a navy base town, lay on the west side of Puget Sound. Its north end, again with the exception of Bremerton, was an elongated extension of peninsular Kitsap and thus doubly isolated. The NEIC was composed of delegations from the small towns and communities of the north end, and was a social organization as well as a "Good Roads Club," founded in 1926. An early January 1927 gathering heard a report on a "Pig Raffle." After the business meeting and refreshments, "all went home Rejoicing." Subsequent meetings focused more on the good work of road district crews, dangerous roads, and private work on public roads. In 1928 one delegation reported an expense of $211.15 on donated labor, teams, and a truck. In 1931 a delegate declared that "all work done" on "Bangor area roads...had been done by their community."[39]

Discussions of low quality roads dominated some meetings, while other problems included the lack of money to build or improve roads in the first place. One county commissioner protested, not against state funding but the unfairness of the apportionment. Kitsap, he pointed out, was the smallest in area among eleven counties of the fourth class, yet "had more miles of road than some of the larger counties and received less gas tax money." Kitsap garnered $131,000 in 1939, but Clallam County at the north end of the Olympic Peninsula received only $1,000 less, despite Clallam's having a mere 475 miles of road to Kitsap's 700 miles.[40]

Apart from illustrating the inherent inequity of road taxes, the commissioner's report to the NEIC revealed how responsive a local, decentralized system could be. At least one commissioner attended many meetings of the NEIC, enabling its delegates to stay in close touch with the elected official most responsible for area road work. Another commissioner, in charge of the northern Kitsap road district, did not show up for some meetings. He was roundly rebuked for his inattention. After he was brought to heel and underwent a transformation, he warned meetings against the "subtle movement toward central government in Olympia" taking over one function at a time, which would result in "a dictator."[41] Curtis wisely avoided the centralization issue when he spoke to the NEIC on October 4, 1933, just after he was elected president of the WSGRA. Instead, in a reprise of his speech to the good roads group, he concentrated on statewide issues. His talk and the clarity of his map and diagram of road "income and expenses" earned him a "vote of thanks" for his "very instructive" presentation.[42]

The NEIC was not alone in its commitment to local decisions about road location. The WSGRA got an earful on the subject of county road development versus centralization when Guy M. Balfour, president of the Washington State Association of County Commissioners, stepped before its 1936 convention. He gave no quarter to the centralizers of the WSGRA. "There is no better example of representative control, nor better organized facilities available for the administering of the secondary highways than the county units as we have them today," he asserted. What looked like uncoordinated, piecemeal road building to Curtis was neither. "The local problems in our program deal with our relations with the postmasters in developing mail routes; with local school boards in developing bus routes; with milk and egg truck operations for best reaching the farms," and with other necessities including "providing needed labor to many of our local people." These activities could "never be replaced by a long distance administration," therefore "the centralization of authority and administration at Olympia," would "be a most serious mistake."[43]

Balfour warned against any reallocation away from the counties, where "the soil…the source of all created wealth" depended on roads to deliver its abundance to urbanites. Promoting tourism and paved primary roads was laudable but not "at the expense of our rural highways." Counties could not build these essential highways from their own resources because state law limited them to a three-mill levy against real property, a hopelessly inadequate assessment. The only solution was to continue the county subvention, meager though it was in the face of growing rural demands for road extension and improvement.[44]

The WSGRA audience could hardly expect the next speaker to propose an increase in primary highway funding, although he did acknowledge "That the State Highway Department should have at least as much money as it has at the present time." Instead, Herbert Horrocks, president of the Association of Washington Cities, wanted to take money away from the counties and give it to the cities. The 1935 legislature, he noted, apportioned about 46.9 percent of net road revenue to the counties, 41.6 percent to the state highway department, and a mere 9.7 percent to the more than one hundred cities in Washington. Horrocks asked the WSGRA to consider the plight of the city dweller who "has to dig down in his jeans" and pay for his city's highway improvements as well as for 75 percent of "the maintenance and construction of the whole county system." Such exactions left the cities "bankrupt." He enumerated city services including parks, playgrounds, and libraries that they somehow had to fund, drains on municipal budgets not experienced by the counties. Horrocks pointed to a BPR survey which apportioned the ideal division of highway funds based on an analysis of tax payments and highway use. The survey "reaches the conclusion that 32.6 of the gasoline tax money should be spent on city streets, 48.2 on state highways, 12.1 on county roads," and 7.1 percent on roads on federal land. Horrocks's stance might possibly endear him to the WSGRA but would hardly comfort Balfour, the head of the county organization. If Horrocks had his way, county funds would fall almost 35 percent of the total state outlay.[45]

In the face of these conflicting demands, Curtis's belief in the complete centralization of highway funding and construction might have been practical had he provided answers to two problems. First, he would have had to forge an acceptable compromise on easing the law's high standards for county roads. It was useless for him to complain about paved, lightly traveled King County roads while heavily traveled state roads went unpaved, unless he were willing to find a way to change the law. His announced solution, taking the money away from the counties, probably would have worked just as well, but the counties would not have surrendered so vital a source of funding and local control without an all-out fight, making his preference a political impossibility. Second, he would have had to craft some way to meet the objections of rural residents who would otherwise lose their influence over the road building activities of their county commissioners. As Balfour told the WSGRA, and as the delegates to NEIC meetings well knew, county commissioners listened closely to the road needs of their constituents. Curtis showed little understanding of this intensely local relationship. He was unlikely to be involved in building a bridge between the highway department in Olympia and, say, the citizens of the upper Kitsap peninsula who fretted about their skimpy road budget. In the last analysis he showed little interest in the practicalities involved in state control of all highway funding.

Nor was he friendly to funding highway programs in cities based on the state road mileage within their limits. The money, he announced in his 1934 WSGRA presidential address, should be allocated "on use, not where purchased. The tank is filled at the city station but in a few minutes the car is miles away, often beyond the county of purchase."[46] He grounded his claim that all vehicles whirled away from urban gas pumps to the far reaches of Washington on the heavy use of state primary roads. Sixty-six percent of traffic was on the less than 4,000 miles of primary highway while the other 34 percent traveled over 40,000 miles of county and secondary roads. It was a weak argument. His figures demonstrated only that primary highways received much more traffic than other roads, not that drivers filling at city stations immediately contributed to the heavy state highway traffic. They may not have, in fact, rolled over any highway. The legislature recognized the problem when it allowed rebates to businesses able to prove "non-highway uses of gasoline" beginning in 1923. On the other hand, many private, non-business urban trips involved no highway use, so the gas taxes paid for driving over town or city streets were not "user" fees but were cross subsidies from non-users to users. Therefore the highways were paid for not "only by those who use them," in Curtis's words, but also by those who did not use them. Curtis could have argued that good roads benefitted all citizens, whatever their direct contributions to highway improvements, but that argument would have warred with the concept of a "user" tax, a belief too fundamental to disturb.[47]

Washington's pay-as-you-go policy engaged Curtis almost as much as the division of funds among the state highway department, the counties, and the cities. The policy limited each year's highway maintenance, relocation, paving, and new construction to the new receipts, less collection costs, in the motor vehicle fund. Because the gasoline tax made up the major part of the fund, people concerned with roads often referred to the whole fund by the shorthand expression "gas tax." Pay-as-you-go contrasted with funding in other states, states that sold bonds to finance highway programs. Speakers at the WSGRA annual conventions—including Curtis—praised pay-as-you-go because it saved Washington from carrying a load of bonded debt. The debt could be so heavy that it drained money away from highway improvements into funding interest and principal. In 1932 President Fred W. Hastings told the WSGRA about the problems of North Carolina, which built its roads from an "excessive state bonding program" but was "having difficulty in meeting the bond payments. We are fortunate in that respect, as we have none."

The next year Curtis reminded his listeners that all WSGRA conventions "but one" had endorsed pay-as-you-go. The result was a state highway system "free from debt," unlike those of North Carolina and "our nearest neighbor, Oregon," which was spending "over $750 per mile of State road for bond interest and retirement and they have no funds for construction." The pay-as-you-go policy was a distinct advantage during hard times, Curtis noted in his 1934 presidential address, because "Washington was one of the few states that was on a cash basis when the present depression came." One positive result was "work for thousands of men, who otherwise would have joined the long bread lines."[48]

Curtis was significantly less enthusiastic about pay-as-you-go in his private correspondence and in his work with the Seattle Chamber of Commerce. He bridled at delays in the proposed loop highway, a road piercing the Cascade Mountains at Snoqualmie and Stevens Passes with gateways at Spokane and Puget Sound. Relying on pay-as-you-go would take too long. He suggested instead an audacious program, dedicating a one-cent increase in the gas tax plus a $8,000,000 bond issue, to a four-year paving program of the 520 miles remaining in 1928, and a nine-year bond repayment period. The loop highway would therefore be a self-contained, self-liquidating program with the one-cent gas tax funding both the paving and the bond retirement. No money from the biennial road appropriation would be required for the loop highway, so all the appropriated funds "could be used on other parts of the State system." His bond proposal went nowhere. The WSGRA approved the loop highway by resolution in 1930 but without his scheme for a combined gas tax increase and bonded debt to pay for it.[49]

Slow progress on paving the loop highway continued to rankle. "My personal opinion is that we are paying a terrible price for our pride" in pay-as-you-go, he wrote. Maintenance was consuming too much of the road appropriation, so the state should be "issuing bonds for permanent improvement."[50] He had no better luck with the bond issue idea at the Seattle Chamber of Commerce. In 1931 he successfully presented a proposal to its resolutions committee requesting the state highway department be granted the power to purchase rights of way wide enough for tree preservation and regrowth in selected areas. But as he noted to a fellow highway beautifier, a "resolution relative to a bond issue failed of consideration."[51]

Curtis leveled one more criticism at over-reliance on pay-as-you-go. Construction of first-class roads resulted in fantastic gains in land values along the rights of way of the new state highways, but none of these increases reverted to the highway fund through taxes. In 1931 he referred to the recently completed four-lane road between Seattle and Tacoma. "This road was located through new territory," and after it was built, "land values increased several hundred percent. Yet the adjacent land owners paid no part of this—receiving the entire benefit of highway construction as a gift."[52] Curtis could have mentioned the higher taxes on the soaring land values, but he and his correspondent understood that those imposts went primarily to county and city coffers. Yet, five years later, no taxes from rising land values were flowing into the motor vehicle fund. This was despite the "very evident" unfairness of abutting property owners reaping the monetary rewards of new or improved roads. "Real Estate people" wanted roads, especially "when they can get all the money for" highways "from someone else…It is easy for them to get the support of community clubs particularly when they can explain that the improvement is to cost them nothing."[53]

Curtis could complain about the unfairness of it all, but the chances of landowners surrendering unearned profits, or counties and cities sharing real estate tax receipts from rising land values, were slim at any time. The chances for such radical tax restructuring in the midst of the Great Depression were less than zero. If Curtis had succeeded in bringing his tax complaint to the table, the property owners, counties, and cities surely would have thrown these words back at him: "I can think of no fairer plan for financing highway work than the one we have in Washington." Curtis spoke those words to the 1933 convention of the WSGRA.[54]

Curtis was similarly ambivalent about the diversion of motor vehicle revenues to uses other than highway maintenance and construction. He was equally uncertain about the unemployed, who benefited from diversion, because diversion, when it came to Washington State, was a welfare imperative of the Great Depression. Curtis recognized the seriousness of the rampant unemployment of the early 1930s, but his sympathies for the unemployed varied, depending on whether the jobless would be helped through road building or some other form of work; direct relief payments, "the dole" in the parlance of the time; or a combination of work and direct relief financed through diversion. He had little use for jobless people who wanted relief without working for it, or for those who depended on relief from money siphoned from the motor vehicle fund. The official position of the WSGRA was clear enough. The second resolution of the 1932 convention opposed "the diversion of motor vehicle funds of the State of Washington to purposes other than road building and maintenance." The president's address of that year reinforced the resolution. "The gasoline tax was not designed to provide funds for propagation of oysters, reforesting burnt-over areas, payment of salaries of teachers, or charity relief to the unemployed in the form of a dole," Fred W. Hastings told the members assembled to hear his talk. Yet those diversions had been proposed in Washington, and the last, "charity relief... in the form of a dole" later passed legal muster before the state supreme court.[55]

Before that date Curtis was in the trenches fighting diversion. Combating diversion was an easy choice for a man whose definition of diversion was extreme, including any deflection of motor vehicle fund revenues away from the state's primary highways to county road building. Most citizens who involved themselves with highways considered the motor vehicle fund to be fungible and were concerned with how it should be apportioned among the state, the counties, and the cities. "I am sorry to see you join with those who are seeking to give more of the State Highway fund back to the counties," he wrote a correspondent in 1931. The counties were already receiving 40 percent of the money, and any "further diversion" would destroy the state's paving program while leaving "us under a heavy expense to maintain a second rate road system."[56]

It was scarcely remarkable, then, that he denounced diversion from the motor vehicle fund to non-highway uses. His private correspondence bristled with attacks on diversion and those who would benefit from it. In September 1931 he wrote about "another effort...to direct considerable sums of money from the State Highway fund. I want to be prepared to renew my fight to prevent any such diversion."[57] In August 1932 he warned of "a serious effort" in the forthcoming state legislative session "to divert the gas tax to feed the unemployed." Meanwhile "many of the King county [sic] unemployed are protesting at being required to work for this money."[58] His alternative was increased federal aid to highway construction, not only for itself, but also to aid "our citizens suffering for food enough to keep them alive—men who are earnestly seeking any form of employment."[59] Briefly, early in 1933, he appeared to favor a bond issue even if gas tax money were diverted to retire the bonds. The unemployment situation was that dire. "We issued billions of dollars in bonds to fight a phantom enemy across the sea," he declared in a jaded reference to World War I. "We have now a real enemy close to home." He assumed that the bond money would be used for road construction and that unemployment relief would be limited to wages paid for road work.[60]

By the time of his 1933 appearance before the WSGRA his assumptions proved unfounded. Curtis hewed to the organization's line while repeating his reservations about indiscriminate aid to the unemployed. After the 1933 legislature passed a $10 million bond issue to be financed by diverting four tenths of one cent of the five cent gas tax, Curtis accused the state of bad faith and of unfairly burdening vehicle drivers. A verbal pledge that "the greater part" of the bond issue and "a written pledge that a considerable part" would be used for work relief had been broken. "Already more than $6,000,000 has been allotted with no funds for State roads and only a small percentage for county." Most of the remaining dollars, probably, would be spent on "direct relief," not on work projects. "The State and the counties are preparing for a large number on the dole and many of the unemployed, desiring not to disappoint them, are planning to be there."[61]

The next year, 1934, saw both the new Democratic governor, Clarence D. Martin, and Curtis address the WSGRA annual convention. Martin was politically unpopular with the good roads group because he had signed the $10 million bond issue to be retired from the gas tax revenue, and because he had authorized bond issue expenditures for "purposes not remotely connected with highways," in Curtis's language. Martin defended the diversion of gas tax money to building various educational and medical institutions. The gas tax funding attracted federal money for construction and saved the citizens of Washington from state, county, or district levies to finance the projects. Construction aside, the diversion raised "the morale of the people, it means more jobs, more activity with their cars…enough perhaps to make up for that four-tenths of a cent per gallon on gas." If the state had not acted to employ people, "I wonder how much gasoline they would have bought!"[62]

Curtis was having none of it. He was incensed over a state supreme court decision allowing bond issue funds for direct or home relief, despite the apparent limit in the legislation to work relief. He asserted that there was "no more justification" for using gas tax money for non-highway purposes than "there would be in taking the interest in school funds and using it for roads." The state functioned merely as a collection agency, gathering money "for the purposes for which collected. Use for other purposes is therefore a breach of trust," he declared in a direct denial of Martin's arguments for the social and economic benefits of diversion. Yet it would be wrong to assume Curtis's indifference to suffering. His lengthy 1933 WSGRA speech, though ostensibly about motor vehicle revenues and expenses, contained a passionate endorsement of highway construction to help those who "by the thousands stand idle, pleading for work, fearful of what the winter will bring to those dependent upon them...I realize that we cannot cure the ills of unemployment solely with public work," but an active program of road building would at least "ease the strain of unemployment...How long would you suffer this to continue if it was your child that was hungry and how long may we expect [the unemployed] to be patient?"[63]

Diversion, unemployment, and pay-as-you-go became less pressing issues after the Franklin D. Roosevelt administration's relief and construction projects diminished the need for diversion and state reemployment efforts. Curtis never was reconciled to the New Deal but its ramped up road spending ended the anguish of the Hartley era for good. By Curtis's death in 1941 the New Deal's emphasis on domestic construction projects, and then on rising defense spending and the military draft, had almost eliminated unemployment among the employable. Meanwhile Curtis focused on other road questions.

Chapter Five

The Good Roads Movement: Nationalizing the Highways

From the 1910s through the 1930s Curtis focused on four issues intertwined with good roads. They were highway beautification; the spectacular rise of trucking and its implications for highway surfaces and road hauling; the Yellowstone Trail Association, especially its extension beyond Minneapolis to Seattle; and the international highway connecting Puget Sound with Alaska through Canada's British Columbia and Yukon Territory. Highway improvement, economic development, and increased tourism were, for him, the elements common to all.

Curtis believed in landscape preservation and improvement along highways. He did what he could to save existing trees, shrubs, and flowers on state roadsides and arranged for new planting, but circumstances limited his arboreal activities. He was deeply involved in other organizations, including the Rainier advisory board and the WSGRA, and in other activities such as his ranch, and the fight over the Olympic National Park. Abandoning or curtailing these responsibilities to pioneer a new movement having limited support was not realistic. Neither state nor county governments were much interested in rights of way along their roads, except for stabilizing them. There were no funds for beautification; a dollar spent on landscaping was a dollar lost to road building. The state controlled relatively little commercial timber on the roadsides, so buying more trees and land was a luxury when construction and maintenance cost so much. Besides, the purchase of a fringe of timber was worse than useless because high winds damaged or downed tree belts only a few hundred feet wide. Swapping private forested land along highways for equivalent public property was difficult when the publicly held tracts rarely were as convenient for the timber company involved. In any case, a timber exchange required a special act of the legislature.[1]

Historian Thomas R. Cox has detailed Curtis's early struggles to save trees, especially roadside trees, work consistent with his labors on behalf of tourism and preservation. No stranger to difficulties or problems, Curtis plunged ahead in 1918 with a speech or article (or both) promoting "protection of the highways," a plea for leaving standing timber along highways in alternate sections of a half mile or so. The expense would total "slightly over five per cent. of the avg. [sic] cost of the road," an amount comforting to him but not necessarily to his readers or listeners. Curtis's effort may have been a part of a larger agitation for saving roadside trees, but in any case it went nowhere.[2]

In 1924 Curtis congratulated the sometimes obstreperous Pierce County commissioner Henry Ball for preserving "big trees" near the new Carbon River Road in Mount Rainier National Park. Curtis's praise came in the midst of his long, frustrating struggle to establish a quality road connection to the west side of Rainier. One tree was "about the prettiest speciman [sic] of the Douglas Fir that I have seen." He was able to "get a picture showing the entire tree," and was sending it to Ball. A rough measurement of the tree showed it to be "about 285 feet high and 155 feet to the firest [sic] limb." Curtis suggested a "bronze plate" nearby with the tree's exact dimensions, the volume of lumber it contained, and other information. There were "larger trees" in Washington State but they were "not accessible to the public," however "this one is so it seems to me important that we have figures available for the thousands who will see it." Whether or not Ball followed Curtis's suggestion, the letter demonstrates his continuing commitment to tree preservation, and to public awareness of the value of trees.[3]

Such random efforts ended in 1930 with the founding of the Washington State Council for the Preservation of Roadside Beauty (WSCPRB). Eula Lee Merrill, the wife of wealthy lumberman Richard D. Merrill, organized the first meeting of the WSCPRB at her Seattle home in March of that year. Curtis was not among the original members but attended the meeting as a representative of the Seattle Chamber of Commerce. The organizers wisely placed him on the constitution and bylaws committee. His presence at the early meetings was sporadic, but his wife Florence attended at least once as the representative of the Woman's Century Club.[4]

The purposes of the WSCPRB dovetailed nicely with Curtis's concerns. The organization promoted highway beautification, generally through the conservation and preservation of trees and plants. It publicized its activities through press releases, letter writing campaigns, and cooperation with like-minded organizations. Its report for 1931-1932 noted various cooperative activities, including rose planting on the Seattle-Tacoma highway with the WSGRA and the Seattle Rose Society. Organizationally, the WSCPRB was an autonomous branch of the National Council for the Protection of Roadside Beauty, itself an outgrowth of the National Committee for the Restriction of Outdoor Advertising.[5]

Curtis's principal role in the WSCPRB was to be a sounding board for Merrill. In 1931 she solicited his advice and suggestions about how she should present a practical application of the ideals of the WSCPRB because she wished to write something useful to one of her acquaintances. Curtis responded with a three-page letter. The idea of a parkway, he wrote, "implies the use of our highways for something besides a way of getting from one spot to another in the least possible time." For that reason the state should control a large area on either side of its highways, to preserve forests, allow the regrowth of cut timber, and prevent stream pollution. Turning to the Mather Parkway (see Chapter Eight), he praised the forest service's "practically ideal conditions and regulations" for control of the section through Rainier National Forest, as well as the "very desirable" improvements to the area within the national park. The unemployment issue loomed large for Curtis, as it did for anyone writing in 1931. Among the "heroic efforts" required to ensure that the unemployed were "housed and fed," he suggested using them to clean up roadsides defiled by logging or highway construction. That suggestion led to two final paragraphs on "the creation of a new forest border," in which he advocated using native materials. Imported trees, plants, and shrubs were a poor investment because they "often" carried insects new to the region and because native plants "in most cases" crowded them out. Sometimes aggressive exotic varieties overwhelmed native species, a situation especially serious with "extremely undesirable" flowers "because they quickly escape from the road to the surrounding fields where they become a pest." Introducing "new varieties of flowers should be under the control of some Department of the State government," he concluded.[6]

Curtis was not shy about delivering advice to Merrill, even when not asked. The man of modest income at least once sent a strong letter to the woman whose Harvard Avenue address marked her as among the city's elite. At issue, Curtis wrote, was a "question of publicity" involving improvements that the WSCPRB wished the state highway department to make. Curtis urged Merrill to talk to the department before going to the newspapers. His statement revealed a deft understanding of the director of highways, Samuel Humes, and of Roland Hartley, who was becoming more cooperative with former enemies as the Great Depression dimmed his political future:

> I think, for instance, if the Director of Highways were to read in the press that somebody was to make him do something, he would not take kindly to the idea while on the other hand, I believe that he would be very glad to discuss these questions with you as president of the [WSCPRB].
>
> I would suggest that you seek and [sic] appointment with Mr. Humes and Hartley for the purpose of a frank discussion of these questions as well as the question of the use of some funds for cleaning up purposes...for the relief of unemployment.[7]

After Curtis pushed preservation and beautification resolutions through the Seattle Chamber of Commerce and the WSGRA, he again urged Merrill to meet with highway officials. They would be receptive, he assured her.[8] Another letter revealed Curtis's respect for Merrill's intelligence and her ability to communicate information. When the Rainer advisory board authorized Curtis to create a committee to study timber preservation along the Mather Parkway, he turned to the board's constituent organizations for names. The Automobile Club of Washington nominated Merrill. Writing to her, Curtis urged her to accept the job "and give us the benefit of your knowledge."[9]

Washington moved tentatively into roadside cleanup and beautification. In 1933 Murrow, the new director of highways, admitted that his department had spent "every dollar available for actual construction" and perhaps had been "somewhat lax in the matter of beautification of the roadsides."[10] Soon the highway department adopted a program of removing "snipe signs" (small roadside advertisements), eliminating refuse and underbrush, and thinning, transplanting, or planting native trees along highway rights of way, a plan developed by the New Deal's Civil Works Administration.[11] It was not until 1937, however, that the department created a consistent, system-wide cleanup and beautification plan.[12] "We are matching 1 1/2% of Federal funds set aside for roadside development,"

Murrow informed a correspondent. The department's work included "the preservation of all natural growth that can be saved," plus "replanting of slopes, rounding of cut slopes for neatness," and similar activities. Before then, Curtis had practically abandoned the WSCPRB, although not roadside preservation and improvement. His lack of interest in the roadside improvement organization resulted from a failed effort to "coordinate" the WSGRA and the WSCPRB. He wanted a good roads committee appointed to "cooperate" with the roadside beautification group, an arrangement evidently planned to develop into a joint committee. His proposal foundered because "each group would resent my including representatives from the other group." Curtis had no doubt where his first loyalty lay. He "let the matter rest" with the WSGRA's endorsement of the WSCPRB's work, "as I did not care to take the [WSGRA] into a controversy."[13] After the state stepped up its highway beautification efforts, Curtis contented himself with advocating recreational improvements and scenic preservation in the Columbia River basin.[14]

Personality conflicts and state activity diminished Curtis's active role in roadside beautification, but there were other reasons. The WSCPRB's highway beautification initiatives were but a minor part of its statewide preservation and conservation efforts; indeed the organization seemed to be as concerned with sign and billboard control and the restoration of denuded roadside landscapes as it was with preservation. The concern was natural enough since logging and road construction hardly created beauty spots, while billboards sometimes blocked views. Curtis preferred to deal with other, more wide-ranging issues. After 1937, when the highway department expanded its cleanup and beautification program, there was less need for agitation of the WSCPRB type. Curtis agreed with the highway department that its principal function was constructing roads, not elaborate, expensive landscaping. "The state needs so much in highways that I would not like to see any large part of our highway money" spent on beautification.[15]

Trucking on highways concerned Curtis more than their beautification. The reason was simple: trucks were destroying highways. As early as 1917 when trucks partly supplanted railroads temporarily paralyzed by wartime freight and troop movements, a symbiosis of truck and road developed. All would have been well had highways been up to the task but they were not, for highways built to handle cars fell apart under the onslaught of trucks. A luxury car of the 1920s and 1930s might weigh two tons or somewhat more if loaded with four adults and their luggage, a burden almost ethereal compared with the weight of a five-ton truck, the behemoth of its day. Five tons was its capacity, not its overall weight when loaded, which could top eight tons. Worse, the monster truck was shod with hard rubber tires that slammed into pavements, cracking them while jarring its drivers and its freight.[16]

The swelling number of trucks and the rise of the trucking industry were truly phenomenal because they impacted so many areas of American life so quickly. Competition, which some observers believed would be controlled by a combination of government regulation and intra-industrial agreement, burst forth in full force. Truckers battled each other for business, unrestrained by trade associations or any serious state or federal control. In the absence of regulations, the competitive situation encouraged truckers to overload their trucks. Competition with railroads threatened the rail carriers because railroads were closely regulated and paid taxes on their extensive property, while truckers paid only gasoline taxes and minimal license fees, and were otherwise practically unrestricted in their operations. At first the railroaders looked on trucking as a boon, much as they first saw the good roads movement. Trucks, they thought, would ease the problem of piecemeal arrivals and departures of goods from urban freight depots to customers. That was true enough, but truckers soon began siphoning off high value freight, carrying it longer and longer distances while undercutting the railroads' tariffs. Worse for the railroads, the federal government continued to regulate them while it stepped in to help the truckers. At the end of World War I the federal government distributed 22,710 surplus trucks to the states, which had only to ask for them and pay their transportation costs. The trucks, useful for road construction, unspent funds from the Federal-Aid Road Act of 1916, a rush of state highway spending, and finally, the Federal Highway Act of 1921 all spurred building roads over which trucks could run, even if destructively. The rail lines could not cut their rates without approval from the federal Interstate Commerce Commission, a circumstance that didn't change with the advent with the advent of unregulated, cut-rate trucking.[17]

Complaints about hard rubber tires and some city ordinances against them inspired tire makers to rush oversized pneumatic tires into production. James Allen, the respected highway engineer whom Hartley dismissed, believed that pneumatic tires would solve the problem of road damage. He "stated that no load can be carried on pneumatic tires great enough to injure pavement." Other experts may have believed in salvation by pneumatic tires. They were wrong.[18]

Urban regulation soon moved to the state level, but state lawmakers faced a bewildering array of regulatory issues. Should trucks be taxed proportionately on weight? Should the tax be on the gross weight of the truck (truck weight plus the rated maximum load tonnage), net weight (the registered carrying capacity), or the tare weight (the vehicle's tonnage standing empty), or some combination of these? What about size? Trucks of the same weight (however defined) varied in their dimensions, depending on the job for which they were designed. What about the uses to which trucks were put? Some trucks were common carriers, like the railroads, accepting freight from all comers. Others operated as contract carriers, limiting their customers to one or a few firms. Still others were private trucks—think of a farmer carrying his crop to market—serving only one individual or company. Still others moved in and out of categories, depending on circumstances. Should all of these uses be taxed and regulated the same or differently? What about one-truck independent operators, were they to fall under the same tax and regulatory systems as the fleet operations? And what about buses?

Once these problems and more were addressed, the difficult question of enforcement loomed. The Bureau of Public Roads (BPR) highlighted some of these issues when, in 1935, it began its comprehensive Highway Planning Survey. The stated purpose was "a survey of future road building" but the BPR's work ranged far beyond construction planning, including checking trucks and buses. L. I. Hewes, the western regional director of the BPR, told the 1938 meeting of the WSGRA about some of the results. "In one state," he announced, "there were three times as many actual 5-ton loads on trucks rated at 1 1/3-ton capacity as were found on the 5-ton trucks." In another state almost 7 percent of buses "weighed empty more than the gross weight license permitted." In yet another state an attempt to create a sliding scale, fees falling proportionately as gross weight increased, was a disaster for the lighter-weight carriers. Four-ton trucks paid $150, while four-ton "semi" or tractor-trailer combinations paid $45. The survey found that four-ton trucks were, on average, underloaded at 3.22 tons, while the trailer rigs were overloaded at 6.4 tons average. "The 4-ton trucks, therefore paid net nearly $46 a ton, whereas the combination paid only about $7 a ton net."[19]

The survey was getting information, not enforcing the laws. That was the job of the states, usually handed to their highway patrols, as in Washington. The Washington patrol's enforcement failures dismayed Curtis, who blamed them on his adversary, Roland Hartley. Hartley underfunded the highway patrol, resulting in "excessive loading" while "the cost of maintenance and repair" far outran "the total monies the Governor is saving." Whether weak enforcement, competitive necessity, carelessness, or all of them explained Curtis's "excessive loading," the reality was that trucks and buses damaged highways. The empirical evidence was overwhelming, because existing highways had been able to handle cars before the coming of trucks and buses, but cracked and crumbled once they arrived. In 1931, long after truck damage became well known, Curtis sent a correspondent two photographs taken six years before of "a section of the Seattle-Everett Highway with practically every joint broken at the corners."[20]

Curtis's solutions, other than better enforcement, heavier pavement, and improved highway drainage, were higher taxes on trucks and buses and reduced permissible loads, along with a suggestion that the heavier vehicles be kept off the secondary highway system.[21] He understood the railroads' and the coastwise shippers' problems, and the unfairness of truck competition as long as truckers paid comparatively little to use publicly maintained roads. The plight of the railroads and the decline of water transport were as empirically verifiable as truck damage to highways, and much commented upon within the WSGRA at the time.[22] Yet as Curtis admitted, there was no dislodging truck and bus traffic. "The commercial use of the highways has come to stay," he told the WSGRA in 1934. Trucks and buses offered "conveniences not furnished by other systems…Boats deliver to docks and trains to depots, while trucks carry direct from producer to consumer." But the traditional carriers had failed to respond to truck competition. Despite his role as a Northern Pacific official photographer, Curtis held to the view that rail and water carriers had their chance to compete and muffed it. "Personal satisfaction with facilities offered to the public and an unwillingness to meet new conditions have cost the older carriers valuable business."[23]

Curtis's enthusiastic response to increased government attention to highways in the 1930s further dampened his interest in any anti-truck crusade. Improved traffic enforcement and effective regulation were partly responsible for his diminished zeal regarding what was a decreasingly relevant single issue. As much to the point was the pro-truck attitude of the powerful head of the BPR, Thomas H. MacDonald, whose beliefs Curtis was more than willing to accept, even if he knew better. MacDonald forcefully denied that trucks harmed highways or hurt railroads. MacDonald's pronouncements, as Stephen B. Goddard proves in *Getting There*, were, at best, misstatements of facts available to the man his worshipful subordinates called "the Chief." Talented and industrious, MacDonald was a first-rate organizer, administrator, and coalition builder. Not given to geniality or friendships, the short, stocky "Chief" controlled highway planning and building through carefully crafted relationships with all people and organizations hoping to benefit by better roads, including roads reconstructed after truck damage. He was not about to criticize anyone interested in highway development, and certainly not trucking companies. For his part Curtis was an avid friend of the BPR for its highway construction in Rainier and in Washington State. Criticism of MacDonald's empire building or public statements was irrelevant to him.[24]

Curtis was far from indifferent to the economic impact of trucking. In 1938 Hewes of the BPR drove home the point to the WSGRA. In the previous ten years (1929-1938) truck registration had increased 77 percent in his eleven-state region. Nationally, he declared that "at least 3,250,000 men" were "employed continuously in driving that fraction of our nation's 4,255,296 motor trucks not on farms" plus buses and taxis. Totting up the total employment related to commercial highway use he found the "truly impressive figure of 6,000,000."[25] By then nobody would have seriously considered interfering with truck and bus travel.

Curtis's commitment to roads resulted in one quixotic involvement, the Yellowstone Trail Association (YTA), one of many named good roads groups promoting a particular route. He stayed with the YTA long after the organization weakened past the point of salvation, perhaps because of his overriding dedication to better highways, but also because he found it difficult to forsake a commitment in the absence of a direct conflict. The YTA, an outgrowth of the good roads movement, was the brainchild of Joseph W. Parmley. Born in 1861 in Wisconsin, Parmley moved west, eventually to the unprepossessing town of Ipswich, South Dakota, in the state's northeast quadrant. He was intelligent and hard-working, with austere features and a drooping mustache, almost a caricature of the small town businessman of the late nineteenth and early twentieth centuries. In time he came to own the local bank, a real estate company, an abstract or title company, and the local newspaper, the latter three conveniently located in the same building. He was a member of the state bar as well as of the legislature, first in 1905 and again in 1907.[26]

Good roads were Parmley's passion, leading to his invention of two horse-drawn road graders, each easy to build from crude lumber and simple to operate.[27] More importantly, his passion resulted in the YTA. Parmley dreamed of a transcontinental highway, no innovation itself, but recognized the need to begin with a more modest road. A series of meetings and conventions beginning in 1912 ended that year with the formation of the Twin Cities-Aberdeen-Yellowstone Park Trail Association, Aberdeen being the biggest town in the Ipswich area. In 1915 the organization changed to its informal name, the Yellowstone Trail Association, though it would not be incorporated until three years later. In 1915 it also adopted a motto used unofficially from the beginning, "A Good Road from Plymouth Rock to Puget Sound." Meanwhile the YTA developed the panoply of activities and structures associated with the burgeoning number of named road associations: officers, an executive committee, regular meetings, a dues structure, auto caravans, and endurance-speed runs reminiscent of earlier coast-to-coast slogs over even worse roads than those of 1915. The YTA instituted publications including route books and maps, newspaper publicity, "Trailmen" and other designated tourist guides at various points along the way, careful route selection, and a successful campaign to admit cars to Yellowstone National Park. It adopted a route marker, a black arrow pointing toward Yellowstone on a chrome yellow circle with the words "Yellowstone Trail" in black capital letters around the yellow band. There were many variations, including large dabs of yellow paint on rocks and poles.

The one thing that the YTA did not do was build roads, apart from a few promotional, local volunteer efforts. Its job, as with other highway associations, was to persuade county and state highway executives to improve its selected route. It welcomed federal highway funding, but its big push was getting local governments and state highway departments to build first class roads according to the best standards of the day. Practically all of its money went for guides, posters, maps, markers, a newsletter, and paid staff. Most of its funds came from cities and towns along the route, usually from their chambers of commerce or other organizations. Individual memberships, donations from dedicated people such as Parmley, some government appropriations, and advertisements in guidebooks made up the rest of the income. Fund raising usually brought in enough money to keep the YTA wavering between small deficits and tiny surpluses, and to move the headquarters from Aberdeen to Minneapolis. The big city was a more prestigious address, and better located as the Yellowstone Trail reached, as the slogan had it, from Plymouth Rock to Puget Sound.[28]

In 1915, a busy year for the YTA, the association marked a "trail" from western Montana across Idaho, and through Washington to Seattle, where the WSGRA, the automobile club, the chamber of commerce, and other organizations had prepared the ground well. All the groups shared the goals of the YTA: building a transcontinental highway or highways, promoting tourism and economic development, and creating an interstate road grid throughout the country. Curtis may have joined the YTA then, for he certainly believed in its purposes, especially tourist promotion and economic development. His surviving correspondence reveals no connection with the YTA until late 1924, when he was a member of the executive committee and working on the problem of delinquent dues.[29]

Curtis might well work on the dues problem, for the YTA skated on thin financial ice even during its most prosperous years. A brief look at the Lincoln Highway Association (LHA), the most successful and prominent of the dozens of named highway organizations, will show just how penurious was the YTA. Carl Fisher, racer, daredevil, visionary, and millionaire, began promoting his "Coast-to-Coast Rock Highway" in 1912, about the time that the YTA was organized. He solicited pledges for coast-to-coast highway materials from auto makers and auto accessory firms for a road from New York to San Francisco. Governments along the route, mostly county governments then, would supply the labor and machinery. The "rock" part of his highway's name stood for a gravel road, a reasonable notion when Portland cement and petroleum-based products were new to rural highways.[30]

Fisher's scheme hit a bad bump when Henry Ford refused to go along, believing that the public should tax itself for all highway expenses. But if Fisher lost Ford, he gained Henry B. Joy, the president of the Packard Motor Car Company, a builder of expensive cars. Joy suggested naming the road in honor of Abraham Lincoln. The name stuck. Even before then Frank A. Seiberling, who had built the Goodyear Tire and Rubber Company on borrowed money and a disused warehouse, was on board. Roy Chapin, the president of the Hudson Motor Car Company, and other important people were there on July 1, 1913, for the official formation of the LHA. Contrast this list of luminaries with Joseph Parmley, successful in his way but fundamentally a big frog in the miniscule pond of Ipswich. Indeed the biggest automotive name on the YTA's executive committee, and a vice president, was John N. Willys, president of the Willys Motor Company, later Willys-Overland, a producer of the world-famous Jeep. There is no evidence that Willys did more than lend his name to the YTA, probably in return for the YTA routing the Yellowstone Trail through Toledo, Ohio, his headquarters. When the time came to fork over real money, his funding went to the LHA.[31]

The LHA had other advantages over the YTA. Named for a revered president, at least above the Mason-Dixon Line, it ran south of the YTA through a more populated, more productive country, except for some desolate western stretches. In 1915 it achieved Fisher's goal, a route (on paper, at least) to San Francisco in time for that city's Panama-Pacific Exposition. In places the road was a seasonally passable track at best, but the same could be said for stretches of the Yellowstone Trail.[32]

The greatest difference was this: the LHA had money and the YTA did not. Five hundred dollars was the largest single individual contribution that the YTA ever received. Contrast this with the funds provided for the LHA's first "seedling mile" in 1914 near De Kalb, Illinois. The Lehigh Portland Cement Company, private citizens, the county commissioners, and the state of Illinois all chipped in. The penurious YTA, by contrast, mustered a mere $4,251.89 from donors in the state of Montana, then the most generous supporter of the YTA, during the 1916-1917 biennium. That was less cash than the LHA raised for a single mile of ten-foot wide demonstration highway. LHA spent almost $1,000,000 from 1914 to 1919 on "seedling" roads, which it encouraged the public to widen and lengthen, while it donated about $218,000 for western highways. Twenty thousand dollars for the LHA's western work came from Willys-Overland.[33]

The YTA usually had just enough in its coffers for a respectable program of publicity, printing, folder distribution, signs, modest salaries, and encouraging public road improvement along its route. About 1925 it began a five year slide into insolvency. Meanwhile its formal as well as informal administrative structures were evolving in ways directly affecting Curtis. While the membership could vote on major matters, such as the route relocation through Washington, discussed later, most decisions were the product of fewer minds. A 1925 letterhead listed an executive committee of seventeen, including Parmely and Curtis, with the easternmost member located in Cleveland, but it is doubtful whether a group of seventeen, strung along a route from Seattle to Cleveland, exerted much influence. At the time, the post office, the telegraph, and the telephone, at high long-distance rates, were the only means of communication. The executive staff, separate from the executive committee, consisted of four people: the president, Curtis as the only vice president, the general manager, and the assistant general manager. Here is probably where the real power and control lay although that cannot be asserted categorically, because few YTA records survive.[34]

What is not in doubt is the influence of Harold O. Cooley, the general manager. "H. O. Cooley," as he liked to sign his correspondence, first jointed the YTA as its unpaid manager of publicity in 1914, though he was on a salary two years later. In some respects he was an ideal general manager. His appearance was nonthreatening. His long oval face was set with small eyes, a long, wide nose, and a wide mouth. It could be described as open and honest, or perhaps, horsey. Intensely committed to the YTA, he was an inveterate traveler over the Yellowstone Trail, tireless, and a speaker whose talks were "filled with facts wrapped up in humorous tales." His "very entertaining" presentations leavened his lengthy discourses, which would have been barely tolerable from a dull speaker. Unfortunately Cooley exhibited less endearing traits. His intensity bordered on a fanatical devotion to the YTA, an impassioned involvement few others shared. His private correspondence at times lacked tact, in contrast to his "funny" public appearances and his optimistic publicity. As he gathered managerial functions into his hands at the central office in Minneapolis, his decisions ventured into policy making, a situation that most members appeared to accept.[35]

Curtis admired Cooley's relentless activities, writing in 1927 that YTA supporters in Seattle believed "the Yellowstone Trail Association" to be "practically H. O. Cooley" himself.[36] Admiring Cooley and accepting the vice presidency in 1924 meant that Curtis took the YTA seriously, yet he was not blind to its problems. He focused on two issues besides the perennial one of raising money: a route change and its consequences, consequences both personal and organizational, and the future of the YTA itself. Together they revealed that the YTA's difficulties antedated its demise in 1930.

In 1925 the YTA membership voted to change its 1915 route, formalized in 1916, from Spokane to Seattle. At first the Yellowstone Trail ran from Spokane southwesterly to Walla Walla, then northwesterly through Snoqualmie Pass to Seattle. The section from Spokane to Walla Walla followed the Mullan Road, originally a military road from Fort Benton, Montana, to Fort Walla Walla, later improved for civilian use during the Montana gold rush.[37] The route was roundabout, compared to the more direct line over Washington's Sunset Highway (the name was an official highway department designation) and requires some explanation. The familiar Mullan Road (approximately today's U.S. Highway 195, state routes 26 and 127, and U.S. Highway 12) was partly the reason, but the Sunset Highway was the other, for it belied its romantic name. Its advantage was its more direct route (following approximately today's Interstate 90) between Spokane and Seattle. The Sunset Highway was shorter by 152 miles, an important consideration when a good day's travel over western roads might be 150 or 200 miles, other things being equal. But other things were not equal. The Yellowstone Trail guidebook for 1916 declared its choice of the old Mullan route to be "a much better road with all the drawbacks of the [Sunset Highway] eliminated." An independent guide of 1915 affirmed the difficulties: "probably less favorable road conditions," "not quite so many large towns," and "more sand is encountered." The Sunset Highway climbed the Cascade Mountains at either Snoqualmie Pass or Blewett Pass, depending upon the year, but neither mountain crossing was easy at first. The old Mullan track was, moreover, the "Scenic Route" through the Palouse, an amazing landscape of buttes and bluffs formed of loess soil.[38] Below the steep hills the loam supported a rich growth of wheat and other grains. After Walla Walla, the scenery to Ellensburg (following approximately present Interstate 84) in the Cascade foothills, was insipid in places. From Ellensburg onward, the foothills and peaks of the Cascade Range and the breathtaking climbs to and descents from the passes more than compensated for drab scenery elsewhere. Passes aside, however, the road from Walla Walla to Seattle was just not that good.

Another reason for choosing Walla Walla was its nearby "junction with the Columbia River Highway," Hill's marvelous Oregon road down the south bank of the Columbia to Portland.[39] In theory the southwesterly dip to Walla Walla opened two major cities, Portland and Seattle, to the tourist, thereby increasing the attractions of the Seattle-bound Yellowstone Trail. In practice it was easier for tourists to choose Portland over Seattle, either by taking the superior Oregon highway, or, once arrived in Portland, by ignoring the trip north to Seattle.

Curtis deplored the YTA's choice of the Mullan route to Walla Walla because it encouraged tourists to avoid the mediocre road to Seattle in favor of the easy, scenic trip to Portland, and perhaps south to California. There were two ways out of this dilemma, according to him, and he did his best to promote them. One way was to improve the road to Seattle. The other was propaganda, to make "a strenuous effort…to offset the work of California and Oregon in maintaining an information bureau at the western gateway at Yellowstone, for the purpose of diverting travel from us." Apparently a 1922 YTA "information booth at Walla Walla to try and hold travel headed westward on the Yellowstone Trail" had not been effective. In February 1925 Curtis was in Minneapolis for a meeting of the YTA executive committee, when he delivered a radio address over station WCCO. Radio was then the novel, advanced way of reaching people with the leisure and money for long trips by car. He praised the products and scenery of Washington, and dilated on the satisfactions of seeing the state by its "splendid" roads. "Take the Yellowstone Trail and follow the sign of the black arrow on a yellow field," he advised. He was dissembling and he knew it, but something had to be done to counter the tourist drain to Portland.[40]

There was another escape, however. It was to change the route, a solution that Curtis dared not mention, because the existing Yellowstone Trail (now Interstate 84) ran through Grandview near his farm. Not only did he wish to keep the paving, economic benefits, and convenience of a well-publicized highway for his farm, he also wanted to extend Seattle's hinterland to the southeastern part of the state.[41]

The issue of tourist traffic diversion continued, despite Curtis's efforts. According to Cooley, the YTA and Seattle, probably represented by the chamber of commerce, reached an informal agreement to abandon the Spokane-Walla Walla-Seattle route to counter the "diversion of travel" to Portland. All parties appear to have agreed that another community would make the formal request for a change to the once-rejected Sunset Highway. Early in 1925 the Wenatchee Chamber of Commerce petitioned the YTA for the change. The Sunset Highway ran, not surprisingly, through Wenatchee, climbed Blewett Pass, and reached Puget Sound, meeting the north-south Pacific Highway north of Seattle or, alternately, at Seattle after a ferry ride across Lake Washington. The Sunset Highway route would effectively end the tourist competition with Portland unless a tourist party had already decided to travel to the Oregon metropolis. One estimate of the loss to Seattle was that only 25 percent of travelers to Walla Walla turned northwest to the Puget Sound city. Curtis's calculation was less favorable to Seattle, 20 percent. There were other issues, however. Even Curtis believed that the Sunset Highway was in "bad condition," a statement he made well after the change was completed, the road was somewhat improved, and he had nothing to gain by belittling the Sunset.[42] In addition, changing the route was chancy because it would involve the expense of moving signs and the loss of memberships along the original Trail. An offsetting increase in memberships on the new route could not be assumed, though communities along it may have purchased memberships in advance, to influence the vote.

Whatever the case, the executive committee agreed to the change, then turned to the membership for a ratifying vote to be tallied on May 8. The vote was decisive, 6,127 for, only 2,133 against. Curtis cast a futile negative vote. The new Yellowstone Trail highway itself received a big boost in 1927 when the Washington legislature passed and the governor approved the then huge sum of $2,439,000 to improve the Seattle-Wenatchee segment.[43] In 1929 Curtis proposed a Loop Highway Association with the object of paving both the Spokane-Seattle and the Spokane-Walla Walla-Seattle roads. The "Washington Paved Loop Highway Association" was established the next year. Though riven by personality clashes, community rivalries, and sectional disagreements, it was another organized voice for improving the old and the new Yellowstone Trail highways.[44]

The highway improvements were desirable, but the route shift undermined the financial health of the YTA, already struggling to meet greatly increased expenses. The association canceled $2,385 in memberships along the abandoned route, and spent about $2,500 to move Trail markers to the Sunset Highway, together almost one-seventh of its 1925 budget.[45] Despite the YTA's route change in favor of Seattle, the city returned just ten dollars to the trail organization. Cooley blamed the Seattle Chamber of Commerce for its failure to appropriate funds for that year, while the chamber blamed Cooley. The correspondence on both sides was not as temperate as it could have been.[46] Curtis would be drawn into this dispute, but independently of the YTA-chamber squabble; he was becoming disillusioned with the YTA. The route change removed his ranch and its area from the trail. He was tired of the YTA vice presidency but could not find a permanent replacement. As early as October 1925 he was questioning whether the YTA should survive west of Yellowstone Park, writing that he had found little interest in the western segment at the YTA's annual meeting.[47]

Survival of the western extension certainly was on Cooley's mind. In October 1926 he threatened the Seattle Chamber of Commerce with suspending operations west of Yellowstone. The Wenatchee Chamber of Commerce's delegate to the annual meeting accepted the termination, but the Seattle chamber urged Cooley to continue full operation. He did so, though cooperation with the chamber seemed to be a one-way street.[48] Curtis tried to get the money Cooley insisted that the chamber had pledged to the YTA, but at first he could not. "I have lost my effectiveness on this particular matter," he informed the YTA's general manager. Embarrassed to be caught in the middle between two organizations claiming his allegiance, he asked Cooley not to mention his letter, writing that "perhaps the best thing to do with it is to read it and file it in the furnace." Curtis's exasperation did not prevent him from accompanying the YTA executive on a fund-raising journey along the western section of the Trail late in 1926. Explaining their modest success, Curtis wrote to another member of the executive committee about some communities along the new route that refused to support the YTA. The trail was there, they said, and would have to continue whether or not they contributed. The experience left Curtis even more inclined to end the YTA's activities west of Yellowstone.[49]

Cooley's fund raising was less than optimal. He hired at least one incompetent solicitor, but he probably had to take what talent he could get, and could not have been expected to carry the fund-raising burden alone. Curtis recognized as much when he praised the "splendid" work of the YTA in its "propaganda for road development and for tourist travel" while he deplored the refusal of "communities that have profited by its operations" to donate money to the trail association.[50] Cooley's difficulties escalated when Parmley, the YTA founder, emerged from quiescence with a "sudden and rather violent eruption" against the general manager. He accused Cooley of mismanagement, carrying too large a debt, refusal to let him, Parmley, review the YTA's ledgers, and failure to order an independent audit of the books. Cooley maintained that neither Parmley nor anyone else in the YTA had the authority to dismiss him unless they bought out his contract (he was writing in May 1927), a contract "which is in force to March 1, 1929." He would do what he could, canceling a debt the YTA owed him for office furniture and reducing his own salary from $6,000 to $4,500.[51] For Curtis, the heated exchanges within the YTA intensified the indifference without. He did not believe that the YTA could survive, "not meeting its salary obligations" especially because "there appears to be a growing lack of interest in all trail associations or similar organizations."[52]

The well-financed Lincoln Highway Association understood the "growing lack of interest" as well as Curtis did. More viable and more accomplished than the YTA, it nevertheless put itself out of business by concluding all "active and aggressive operations" at the end of 1927. For a few more years the board of directors met, but its last major undertaking was to memorialize Abraham Lincoln with about 3,000 markers along its old route. Local Boy Scout troops placed the ungainly markers, stubby cement obelisks set astride longer, narrower cement bases on September 1, 1928. The LHA was effectively at an end.[53]

Cooley did not get the message. But what was the message? It was a tectonic shift from specific, organized, private propaganda for particular highways to a national vision of a rationalized network of roads, bureaucratically controlled in its construction and maintenance, and, by previous standards, lavishly funded by cooperative federal and state governments. The federal highway act of 1921 called for establishing nationwide interstate routes, in effect mandating some uniform system. What clinched the change was the 1925 numerical system suggested by the American Association of State Highway Officials to end the chaos of 250 named highways, if not as many active associations. A Joint Board of the AASHO and the BPR agreed to mark the national highways with even numbers for roads running east to west and odd numbers for those wending north to south. The numbering of east-west roads began at the north in increments of ten, with U.S. 10 in places skirting close to Canada, while U.S. 90 traversed deep Dixie. U.S. 1 hugged the east coast, U.S. 101 the west coast. States could number or otherwise designate their routes as they chose. At the same time the Joint Board picked the highway marker, black numerals on a white shield, and adopted a uniform system of information signs. Finally, the board decreed that the private trail markers would have to come down to avoid overwhelming the motorist with what was already, in some places, a confusing jumble of overlapping private markers. The secretary of agriculture approved the program in late 1926. The federal signs went up, the private markers came down where they could be readily removed, and the day of the road associations passed.[54]

Nor was federal-state cooperation for a comprehensive highway system the only development militating against the fading trails. Private travel agencies were committed to providing customers with the best routes, not specific ones. Independent travel guides had existed for years but became more elaborate. The regional guides published by the American Automobile Association (AAA) for instance, listed approved accommodations and tourist attractions. Oil companies began providing excellent free highway maps at their filling stations, maps not confined to particular roads. The AAA and some insurance companies provided towing service. The AAA established offices in major cities and some not so major, supplanting the roadside assistance of the Yellowstone Trail.[55]

The next question is, why didn't Cooley get the message? The first and most important reason was the continued income and employment of Harold O. Cooley. His resignation or unilateral termination of the YTA would have ended his claim on his salary that extended into early 1929. Leaving in 1927 would have been a capitulation to Parmley. His "plan and dream" was to clear the YTA's debt. Besides, he wrote to Curtis, "I love the Yellowstone Trail and its work."[56]

Meanwhile, Cooley's optimism rose and fell with the YTA's income. Collections in 1928 were adequate if spotty, but by mid-1929 Cooley was in despair, having a scant $1.98 cash on hand as of June 1.[57] "The Trail is the greatest auto travel asset that the nation has," he declared with his characteristic hyperbole, unable to understand why it was not better supported. He asked Curtis what he would do under the circumstances. Curtis replied that he would wind up the affairs of the YTA at once, explain to the members what was owed, and hope for funds to cover the debt. He was disappointed, he repeated, in the reactions of the commercial organizations along the trail, including the response of the Seattle Chamber of Commerce.[58] Cooley did not take Curtis's advice, but instead canvassed the trail in the fall and winter of 1929 with two solicitors and "succeeded in getting a substantial sum of money."[59] Heartened by this turnabout, Cooley delivered a rousing speech in Aberdeen at the end of 1929, and was making plans for 1930.[60]

Cooley's optimism was short-lived. By mid-March 1930 he was out of money. A debt he calculated at $3,976.27 in 1927 had ballooned to $6,176.45. A solicitor charged with refilling the YTA's coffers had, he strongly suggested, absconded with whatever funds were collected. He sent similar letters to Curtis and to A. J. Dahlman of Lemmon, South Dakota, the last president, announcing that he was closing the office and removing some personal effects. His antagonist, Parmley, gathered the YTA's "records, correspondence, and minor items," and sent them to Ipswich. After they arrived, Parmley made a "hasty examination" of them, finding "nothing of value therein," perhaps because they contained so little about his activities and so much about Cooley's valiant if sometimes misdirected efforts to salvage the YTA. There is no doubt that Parmley destroyed the material.[61] Soon he founded a new trail association which existed, modestly, through two incarnations but effectively ended with his death in 1940.[62]

What, then, of the Yellowstone Trail Association? One salient fact about the YTA and its sister organizations is their brevity, for their heyday lasted only about fifteen years. An equally remarkable fact about them is how much they accomplished in that time. They raised the public awareness of the need for durably surfaced roads, marked out reasonably reliable continent-spanning highways, promoted tourism, and with expanding tourism, a growing appreciation of the country's dramatic scenery and of its continental breadth. They worked hard to improve the roads along their chosen routes. They succeeded, although this activity, as with the others, is impossible to quantify. By 1925 the some 96,000 miles of federal highways were appallingly backward by later standards but greatly superior to their predecessors (where predecessors existed at all) in paving, curvature, grades, sight lines, width, drainage, and roadside services. The YTA did its share in bringing about all these changes.[63]

For Curtis the results were less positive, indeed probably caused him more grief than joy. The 1925 route change took the Yellowstone Trail away from his personal Grandview-Seattle axis. Seattle remained the Puget Sound terminus of the trail but the refusal of the Seattle Chamber of Commerce to support it wholeheartedly vexed him. He sympathized with Cooley, but thought he lacked judgment at times. During the dispute between the YTA and the Seattle chamber he wrote the YTA general manager that "if I should ever happen to write to you again asking that you write to anybody here, please don't open your letter by stating that you are writing at my request." Cooley's statement made it appear as a personal plea for Curtis and not an entreaty from the YTA. "I rather think that without that one line in your letter I would have had a thousand dollars for you now."[64] Nor was Curtis enthusiastic about the timing of Cooley's fund-raising trips. The "best way," to raise money was "to go over the ground early in the spring," a trip Cooley did not make every year, perhaps because of thawing roads.[65] Had Cooley taken Curtis's advice the association could have been financially stronger, but without a radical reversal of the circumstances overwhelming it, the YTA was doomed. Its demise freed Curtis to embrace a new road enthusiasm, an international highway from Seattle to Fairbanks, Alaska.

Donald MacDonald, a somewhat eccentric engineer, received credit for the idea of a highway between Alaska and the rest of the United States. MacDonald, no relation to the head of the BPR, worked for the Alaska Road Commission, a quasi-military agency of the War Department. The commission, an active and visible federal presence in Alaska, adopted MacDonald's idea; so did a panoply of other organizations. Business people in Fairbanks soon established an International Highway Association, as did their counterparts in Dawson, the capital of Canada's Yukon Territory. The Alaska legislature asked Congress to advance the idea and appropriated money to propagandize the project. Practically every chamber of commerce in Alaska endorsed the concept, as did Seattle's. The WSGRA urged intergovernmental cooperation to extend the highway from Alaska to Patagonia. Other Seattle and Washington organizations chimed in, as did several national groups including the American Road-Builders Association and the United States Chamber of Commerce. The Department of the Interior became interested, while representatives of the Department of State held meetings with their Canadian peers. In 1930 Congress appropriated $10,000 for a commission to study the route. A 1931 joint meeting with the commission's Canadian counterparts concluded "that the project was feasible from an engineering and constructional standpoint."[66]

If all this seemed to forecast a done deal, it didn't. Alaska was vast, politically a territory with a miniscule, mostly coastal population, and with only one sizeable interior town, Fairbanks, the highway's projected northern terminus. Alaska was remote, effectively a transportation island reachable only by ship or the small-capacity aircraft of the time. Fairbanks was isolated during the winter months, except for the 470-mile federal Alaska Railroad and a 371-mile sled road to the coast. Short summers opened the sled road, the Richardson Highway, to motor traffic, but stream flooding sometimes tore out bridges or sections of the road. Except for the Steese Highway, 160 miles north from Fairbanks, and five short stretches elsewhere, summer turned much of the interior into a muskeg swamp. "Bush" airplanes operating out of Fairbanks were the only dependable connection to some isolated hamlets, except for the three to four months when the Yukon River and its tributaries were open to shallow-draft navigation. All of Alaska's industries were extractive and most were seasonal. In the early 1930s they consisted mainly of "fish, gold, copper, and furs." Southern coastal Alaska was generally temperate enough, or at least not as cold as the "Seward's Icebox" cliché would have it, but even in summer it could be chilly, windy, rainy, and overcast. Interior weather was another matter. During the almost incredibly mild winter of 1929, the low temperature in Fairbanks was -38° F, while the summertime high was a more normal 86° F.[67]

There were other problems. Alaska's 590,000 square miles contained more than twice the land area of Texas and included several varied climatic zones. Its tiny population of sixty thousand, about half of it Native, was unlikely to stir much sustained interest in Washington, DC. Its lone political emissary was a non-voting delegate to the House of Representatives. There was no representation in the Senate. The territorial governor was not elected by Alaskans but was a political appointee. Governors came and went with changes in the national administration, if not sooner. The limited powers of the legislature, the small and scattered population, Alaska's modest taxable wealth, and the hundreds of miles separating many small settlements meant that federal funding supported the territory, some seven million dollars annually versus territorial levies totaling about a million and a quarter.

The separation of Alaska from the rest of the United States required any road to be built through Canada's British Columbia province, raising questions of international cooperation.[68] The most direct route was from the Washington-British Columbia border to the southern tip of southeastern Alaska, but Alaska's narrow panhandle was inhospitable to roads. Steeply rising mountains, rushing rivers, wide fjords, and several glaciers rendered it impassible. The highway would have to run from southern to northern British Columbia, east of the forbidding coast range, through a province sparsely settled in most of its northern half. For political reasons the route did not go directly to Fairbanks but detoured to Dawson, then the capital of the Yukon Territory, a thrust adding about 100 miles to the project despite the Yukon's barren vastness. If the phrase "sparsely settled" applied to Alaska and northern British Columbia, then "almost deserted" would describe the Yukon. The sprawling territory of 207,000 square miles contained a permanent population of 6,000 at most.[69] In sum, the enormous land through which the road had to be built was in many places a population void. The land itself could not afford the cost. The people who lived there could petition for a highway all they wished but they had practically no political influence where it counted. Even some private resolutions in favor of the road called for a thorough investigation of its potential, not its construction. The distance from Seattle to Fairbanks was 2,256 miles, a bit more than half of which would be new construction, 1,000 miles of it in British Columbia and the Yukon. The estimated cost was almost $14 million "for a graveled road surface 16 feet wide on a graded roadway 24 feet wide," with ditches, culverts and bridges put in. In other words, the minimum standard of the day.[70]

Yet there were reasons to be enthusiastic about the project. It was visionary, but seventeen years prior, a transcontinental highway was visionary. The premier of British Columbia, Simon Fraser Tolmie, favored the project. If citizens of Alaska and the Yukon lacked clout, the Seattle Chamber of Commerce had some, and the United States Chamber of Commerce and other organizations had more. The Great Depression struck soon after MacDonald revealed his vision, so he added economic relief through road building to his repertoire of reasons for the road. Then there was the perennial reason: the "Last Frontier" beckoned from Alaska. It was a "new" frontier needing only adequate transportation and settlement to blossom and begin sending its products southward. "Where is the logical place to go?" Dr. Evan W. Holway of the American Legion asked the WSGRA. The question was rhetorical. "Why, my friends, turn your eyes northward...We are seeking to expand the frontier...to enlarge the opportunities for happiness and success" by developing "this vast store of natural wealth and resources."[71]

By the time of Holway's 1934 address Curtis was disillusioned with the progress of the international highway, convinced that boosterism alone would do little in the absence of any firm commitment from a government. At first he was excited, working with MacDonald, Malcolm Elliott, the president of the Alaska Road Commission, and others to promote the road, although he cared less for the economies of Alaska and Canada than he did for the prospect of increased tourism. When Tolmie organized an auto caravan from Vancouver to Hazelton, British Columbia, an 815-mile trip one way, Curtis went along as the expedition's unofficial photographer, snapping dozens of photographs and thoroughly enjoying himself.[72] The reception at Hazelton was a love feast. The lieutenant governor of British Columbia, Governor George A. Parks of Alaska, and several other Alaska representatives met the caravan. International joviality and pleasant speech making marked the occasion. After his return, Curtis worked with Ernest Walker Sawyer, of the Interior Department and a member of the expedition, to prepare a commemorative album to present to President Herbert C. Hoover and the members of the commission. About the same time British Columbia completed aerial and ground reconnaissance for a general location of the road. Before March 1931 Curtis developed what one correspondent termed a "very splendid address on the International Highway."[73] Two years later Curtis, by then chairman of the WSGRA committee on the international highway, and others organized a second caravan, an "International Progress Tour" from Seattle. The tour made a briefer excursion into Canada, where the bonhomie of the 1930 trip was repeated. Then the caravan went to Chinook Pass for the dedication of the Mather Parkway (see Chapter Eight). In the intervening two years Curtis kept up a drumbeat of correspondence about publicizing the road.[74]

Unfortunately the international highway never moved far beyond publicity, a relatively inexpensive government activity. Politicians and officials on both sides of the border were willing to appoint the first international commission and a later, similar one, but were chary of spending any real money on the project. The warning signs appeared early, when Curtis, Hill, and Sawyer tried to awaken Governor Hartley's interest. Hartley signed a telegram, drafted by Curtis, to an international highway committee meeting but would not attend. Sawyer wrote Hartley to suggest working closely with British Columbia but received no reply.[75] In 1930 the Canadian federal government handed off the highway to British Columbia. At stops along the way of the 1930 caravan Tolmie announced "on many occasions" the support of the British Columbia government "so far as it could be undertaken without interfering with the regular program for expansion of the provincial road system." In 1932 he admitted to Curtis that the "matter" was not "active at all."[76]

Private initiatives proposed to act where governments provided little more than lip service. The indefatigable Hill proposed forming a syndicate to build the needed miles in return for concessions such as gas stations, a resort, and possibly some of the gas tax receipts. Nothing came of it. Tolmie wrote Curtis about "a proposal from a bunch of Englishmen" also willing to build the road for certain concessions. "They wanted to build a double roadway with a boulevard in the middle and electric lights." Tolmie wondered what the bears and everyone else "way up there in the bush" would think of that. To Tolmie it was all a stock-selling scheme. He discouraged it.[77]

The United States commission released its report in 1933, one of the worst years of the Great Depression. It contained what by then was mostly boiler plate about the untapped riches of Alaska and the Yukon. Its calculations of increased tourism were conservative but scarcely credible in 1933. It failed to inspire any workable plan for building the road.[78] By 1934 Curtis was close to admitting defeat for the international highway. He agreed with a British Columbia correspondent that the province should build roads to reach its existing population. Tolmie, by then a private citizen, stated the case more forcefully. He noted that "the province is responsible for its roads, both construction and maintenance," therefore a road to Alaska "would be an absolute impossibility" unless the Canadian federal government took a hand. Tolmie regretted that federal aid was unlikely, because "it has always been a difficult matter to get the easterners interested in anything west of the [Rocky] mountains." To another Canadian correspondent, Curtis wrote of how he was "very greatly disappointed in the attitude of the city of Vancouver relative to the highway." In 1935 he wrote to Sawyer that British Columbia's attitude had improved and that the province "would welcome a real plan to finance and build the road." He hoped that Sawyer had such a plan. Sawyer, retired with the rest of the Hoover administration after the Democratic sweep of 1932, was unlikely to have any plan or any way to implement it. Curtis noted that a bill sponsored by Anthony Diamond, Alaska's delegate, was "moving along with the support of the President." With the funding withdrawn, however, it was "just another gesture." The Seattle situation was equally discouraging. "Locally the subject has drifted into the hands of a bunch that seems to think that the road can be built with talk."[79]

Continued interest in the highway, while not ardent, was active enough. A second commission appointed under the Franklin D. Roosevelt administration studied the problem, recommending essentially the 1933 commission's thrust east of the coast range, north and west through Canada before heading west to Fairbanks.[80] By then Curtis's interests had shifted to defense highways.[81] Unfortunately for the direct road from Seattle, awareness of the 1933 route spawned competing suggestions. In 1941 enthusiasts on both sides of the border favored a "United States-Canada-Alaska Prairie Highway Association" with the object of promoting a highway beginning in Edmonton, Alberta, far away from the coast. Later the same year an official Canadian report argued for a sort of compromise, an interior road beginning at Prince George, British Columbia, and running north between the coastal mountains and the prairie route. Strategic thinking, nevertheless, was pushing the final decision toward the prairie line. By 1941 Canada had established a Northwest Staging Route, building or improving airfields from Edmonton to Whitehorse and beyond through the Yukon toward Fairbanks. The United States Army Air Corps soon was flying fighter planes up the route, hopping from landing field to landing field on the way to its expanding air bases in Alaska. Therefore the Edmonton-Fairbanks route made sense as a supply road for the airfields. The Japanese naval and air attack on Pearl Harbor in December 1941 lent new urgency to a decision. In March 1942 the governments of the United States and Canada signed an agreement for a road following the staging route from Edmonton (technically the road began at Dawson Creek in British Columbia) through Whitehorse to Fairbanks. Additional reasons for the route were its greater immunity from Japanese attack and the absence of the heavy snows and rains plaguing portions of the Seattle-Vancouver-Dawson-Fairbanks line. Wartime emergency, not northwest economic development, decided the route and the timing of the road to Alaska.[82]

Curtis's shift from optimism to pessimism about the Seattle-Fairbanks route was realistic. The initial excitement in the United States produced one federal study commission, while continuing mild interest brought forth another. There were joint Canadian-United States discussions in between but, beyond them, no sustained action. There were reasonable questions about whether yet another transportation artery would bring any significant new development to a remote area already served by the federal Alaska Railroad from Seward to Fairbanks, the private White Pass & Yukon Railway from Skagway, Alaska, to Whitehorse, seasonal river transportation, small airplanes, and an extensive if fairly primitive road system. When the federal government did embark on a fresh, Depression-inspired program, it moved 201 farm families, mostly from Michigan, Minnesota, and Wisconsin, to Alaska's Matanuska Valley north of Anchorage. The experiment proved to be expensive and marginally successful at best.[83] World War II began for Canada in 1939, ending all interest in a highway venture unrelated to Dominion defense, except for the belated 1941 report. The highway, when it came, was the result of American defense anxieties, not Canadian.

Overall, however, Curtis deserves praise for his relentless struggle for better roads. He worked for good roads through the WSGRA in an era when hard work from private individuals contributed to improved road transportation. It required courage to contend with Governor Roland Hartley's hostility to highway improvements at a time when Hartley's popularity could have persuaded Curtis to keep silent. Yet when Hartley's popularity faded and he embraced highway spending programs in an attempt to revive it, Curtis welcomed the governor's change of heart. Highways were, to Curtis, never ends in themselves, but a means to improve Washington's economy by opening access to markets and improving tourism. Public, bureaucratically directed activity gradually superseded his work for the WSGRA, roadside beautification, and the Yellowstone Trail. Private initiatives were never enough to fund the international highway, only wartime exigencies did that. In retrospect, Curtis struggled manfully to create the situation that rendered his efforts superfluous.

Mountaineering

Curtis took mountaineering very seriously. Here he poses atop a cairn on the summit of Mount Shuksan following his 1906 climb with W. Montelius "Monte" Price and perhaps others. *Washington State Historical Society, Tacoma, 1943.42.7516.*

In 1907 Curtis led a military-style assault on Mount Olympus on the Olympic Peninsula. Here a group of Mountaineers climbs toward the Olympic summits before bad weather forced a retreat. *Washington State Historical Society, Tacoma, 1943.42.43713.*

During the summer of 1917 Curtis was the guide for the Mount Rainier park concessionaire, the Rainier National Park Company. He took fifteen groups of climbers to the summit of the mountain. Here he stands at the left and a little apart from one group. He also conducted tamer tours below the summit, all "without the slightest accident." *Washington State Historical Society, Tacoma, 1943.42.40158.*

MOUNT RAINIER IN PHOTOGRAPHS

Curtis took dozens of photographs of his beloved Mount Rainier, including this one on a summer 1909 outing of the Mountaineers. He titled the photo "Lost to the World," though it's doubtful that he happened across "Jack and Nettleton, Miss," gazing at an ethereal, cloud-wrapped world. The photo is almost certainly posed. *Washington State Historical Society, Tacoma, 1943.42.15686.*

Curtis captured this stunning view of the Cowlitz Glacier in August 1911 as it flowed southeastward from the flanks of Mount Rainier. *Washington State Historical Society, 1943.42.21809.2.*

This remarkably crisp photo captures Mount Rainier from Lake Eunice in August 1935. *Washington State Historical Society, Tacoma, 1943.42.61474.*

The Dancing Girls

In September 1931 Curtis photographed his most whimsical series when he and young women from Seattle's Mary Ann Wells School of the Dance drove to Sunrise (Yakima Park), an eastern section of Mount Rainier National Park newly opened to cars. At least thirteen photos resulted. Here the students are dressed in their dance costumes, but are wearing coats against the September chill. *Washington State Historical Society, Tacoma, 1943.42.58586.*

Symbolism was on less than subtle display in the Sunrise series. Here nine dancers gather around a brand-new 1932 Nash sedan transformed into a chariot, either to pull it forward with silken reins, or to offer psychic encouragement from the rear. *Washington State Historical Society, Tacoma, 1943.42.58587.*

Five dancers removed their slippers to appear barefoot on the margin of Fairy Lake in this imitation of ancient mountain worship. Three of the young women have plaited their hair with flowers. *University of Washington Libraries, Special Collections, 1575.*

Chapter Six

Mount Rainier: Climbing the Mountain, Founding and Leaving the Mountaineers

When Curtis first saw Mount Rainier close up he probably had no idea of the impact that the lofty pile would have on his life. His involvement with the mountain would lead him to climb it many times, to found, with others, one of the leading mountaineering clubs of the West, and to leave the club in a bitter policy dispute. Ironically, the dispute involved, not Mount Rainier, but a proposed national park in the Olympic Mountains. He would go on to preside for many years over an active citizen group promoting and developing Mount Rainier National Park for tourism. Along the way he would form friendships and endure enmities and defeats, but see the realization of some of his dream for Rainier.

Curtis caught "mountain fever" early. In 1896, when he was twenty-one, he walked four days through dense fir, spruce, and hemlock forests east from Tacoma to reach Mount Rainier. He had no money to do otherwise, because he had moved within the year from the family's Port Orchard home to work in his brother Edward's photography studio in Seattle. What is remarkable about the trip, besides Curtis's budding romance with the mountain, is his ability to save enough from his salary to afford any trip at all, and his brother's generosity in giving him two weeks or more away from work.

He did not mention spending the time to hoof it between the two rival cities, so the probability is that he took a train or a Puget Sound boat. To the cost of the round trip ticket he had to add the cost of provisions, even if he avoided the expense of such "inns" as there were east of Tacoma by sleeping outdoors. He could have paid to ride a horse-drawn buckboard stage over a dirt track from Tacoma to the mountain and back again. James Longmire, the rugged, gaunt explorer, mountaineer, guide, farmer, and tourist entrepreneur, had already opened a road of sorts to his rustic mineral bath houses and small, very basic hotel. Even so, a healthy young man like Curtis would prefer walking behind or alongside the stage as it bounced over the roughest parts of the pioneer road.[1] Why should he bother with a stage? Walking the sixty-one miles from the end of the streetcar line to Longmire (now inside the park) avoided additional expense, afforded the solitude he craved, and allowed him to explore the mountain meadows, streams, and valleys without considering the wishes of others.

Curtis made no claim to climbing the mountain. Almost certainly he did not try to reach the summit.[2] The only reason his trip is known at all is because of his brief mention in an extemporaneous aside during his 1934 presidential address to the WSGRA. Then his purpose was not to extol Mount Rainier or rhapsodize his four-day tramp but to contrast his lengthy walk with the 1934 experience, when "over smooth pavements, this same trip takes but two hours." In 1896 "the whole mountain region was mine to enjoy alone," he declared with a small exaggeration. In 1934, he asserted, 250,000 people visited the park, a serious overstatement because the official total for fiscal years 1934 and 1935 together (July 1, 1933-June 30, 1935) was only 151,531. Curtis may be forgiven the misrepresentation because he was trying to buoy spirits during the Great Depression, when travel to the reservation slumped. At least Mount Rainier National Park was an assured portion of the United States' preserved, protected landscape.[3] The same could not be said of the mountain and its surroundings during his first visit in 1896. The Forest Reserve Act of 1891 was little more than five years old, while President Benjamin Harrison's designation of the Pacific Forest Reserve in 1893 was about three and a half years past. The act and the forest reserve were nothing more than gestures. Neither included an appropriation or any enforcement power; a later president could revoke Harrison's designation.[4]

Gestures are significant, nonetheless. Curtis tramped to the mountain at the conclusion of a revolution in American perceptions of wilderness. To the earlier Euro-American mind, wilderness was hostile and forbidding, a place of dangerous beasts and savage people, an unwelcome contrast to settings where human effort converted savagery to settlement; to towns, cities, roads, farms, and industry. Wilderness was to be subdued, then eliminated. The idea that wilderness was repellant and should be destroyed had its detractors in the eighteenth century Romantics, but the importance of wilderness as a repository of spiritual values took hold only in the mid-nineteenth century and after. Ralph Waldo Emerson, Henry David Thoreau, and others were responsible for this reversal of view. To them, nature was fundamental to human development. Destroying nature stunted every person's potential.[5]

Such conceptions belonged, principally, to elite thinkers in the eastern United States. They developed, in part, from the recognition that urbanization and the cities sustained by farming, lumbering, and mining had not created an elevated civilization or particularly civilized people. Mid-century cities, with their propensity for human illness, devastating fires, mob violence, and for the elevation of material gain above other values, were not what the earlier proponents of civilization had envisioned. People, therefore, required an alternative to civilization: untamed expanses where spiritual values could be created or restored in visual, tactile immersion in nature.

There were plenty of people, easterners and westerners, who regarded such beliefs as silly or even dangerous nonsense. Despite those detractors, preservationist thinking undergirded the park movement. The movement's hard work in the real world resulted in Congress's creating, in 1864, the core of the later Yosemite National Park, and, in 1872, the Yellowstone National Park. Snowballing preservation produced national forests, state parks, a vigorous urban parks movement, and what Roderick Frazier Nash described as a "wilderness cult."[6] Organized preservationist activity (for example, the Sierra Club, founded in 1892); the primitivist writings of John Muir and others; vacationing as anodyne to urban-induced "neurasthenia," a generalized bundle of unspecific nervous disorders; and hunting and fishing as restorative of lost or threatened traditional American virtues, combined to quicken interest in wilderness adventure.[7]

When Curtis made his pilgrimage to the mountain, local boosters and commercial interests in Tacoma and Seattle were in the thick of a fight to make a national park out of Mount Rainier and its environs. Booster publications mixed paeans to the natural beauty of the Pacific Northwest and Rainier's sublimity with not-so-subtle references to the potential for tourism and economic development in the Puget Sound region. How much this local output encouraged settlement and tourism is open to question. Historian Theodore Catton correctly observes that support beyond Seattle and Tacoma was essential to creating the national park. Railroad promotion, beginning with the Northern Pacific's arrival in Tacoma in 1883, probably contributed much more to tourist travel and preservation because tourists wanted an unsullied landscape at their destination. The promotion became especially intense after competing lines arrived. Together the Northern Pacific, the Union Pacific (to Portland), the Great Northern (to Seattle), and the Milwaukee Road (Seattle, 1911) advertised the scenic wonders of the Pacific Northwest far and wide. Local mountaineers and naturalists who worried about vandalism, fires, poaching, uncontrolled timber cutting, and rampant exploitation of the tourist trade, joined the cities and the railroads to promote a park. Scientists, especially geologists, added their weight, as did outdoor groups such as the Sierra Club and the Appalachian Mountain Club. The arguments for a park thus clustered around a wide range of desirable outcomes: recreational, scenic, spiritual, scientific, environmental (such as watershed conservation), preservationist, and nationalist, the last impulse borne on the belief in the superiority of American scenery over Europe's.[8]

All or most of these goals were potentially conflicting. For instance, how could tourists enjoy recreational support such as roads, trails, guides, and accommodations if all of the park were to be preserved in its natural state? Official roads, marked trails, designated campsites, sanctioned hotels, and all their adjuncts would have to be provided, if for no other reason than to discourage tourists from running willy-nilly over the landscape, damaging the natural beauty they had come to admire. Even people with a common goal—popular recreation—would raise serious disagreements over the number, location, priority, and construction details of roads. In later years Curtis argued for an interconnected road system, exploiting views while it reached all park entrances, roughly its four corners, and extended to most tourist areas deemed adequate for the accommodation of large groups. An elaborate infrastructure of trails, hotels, and campsites would supplement the roads. His equally dedicated opponents would severely limit roads, partly because roads scarred the primal landscape. Roads, where they were allowed to exist, would funnel the tourist hordes to a few carefully prepared sites while opening vast areas to hiking, horseback riding and mountain climbing only. Then the truly dedicated recreationists would be free from the intrusive racket and odors of horse-drawn conveyances, buses, and automobiles, not to mention the superficial pleasure seekers themselves.

Did the park's proponents fully realize the looming contradictions embedded in their supporting arguments? Did they see potential conflicts with some positions taken by their allies in the park fight? Some advocates probably were so focused on the outcome as to ignore any potential contests over the park's use. Perhaps others naively assumed that unity in their struggle would survive their victory. Some preservationists doubtless knew of their counterparts' successes in Yellowstone and New York's Adirondacks, and assumed that any threats to the wild nature of Washington's national park would be likewise defeated. From 1893 to 1898 the state's senators and representatives responded to these heterogeneous pressures by introducing six bills creating a park. The final effort cleared Congress early in 1899 and President William McKinley signed it into law on March 2 of that year.[9]

The park was small as national parks went, about 320 square miles. Among other incidents critical to its creation, the final bill allowed the Northern Pacific Railroad to exchange original land grant acreage within the reservation boundaries for forested lands in any state through which it ran. The privilege was extended to other property owners, none likely to claim anything approaching the railroad's some 450,000 acres. The park act gave administrative control to the Department of the Interior but appropriated nothing for Rainier's administration, preservation, or improvement. The Secretary of the Interior could promulgate rules and regulations, but their enforcement or any development of the park depended on revenue from leases or concessions, unlikely sources of adequate funds. The upshot was that the five hundred or so annual visitors to the park largely policed themselves until the U. S. Forest Service undertook a rudimentary ranger program of patrol and improvement, beginning in 1903.[10]

The park's centerpiece is Mount Rainier, its 14,411-foot massif dominating the skylines of Seattle and Tacoma, and everything else within one hundred miles or more. It is the fifth highest mountain in the country, but the most impressive. It rises in solitude, aloof from its brothers and sisters in the Cascade Range, glistening with glaciers and compacted snow far above its modest lowlands. It is a volcano, still seething within its white mantle, still sending steam to its summit. Robert Shankland best expressed its diversity when he wrote how "from bottom to top the hiker encounters the plant and animal life of four climate zones: broad fields of flowers growing at the edges of the ice, great coniferous forests, waterfalls and streams running with glacial boulders, alpine meadows gleaming with lakes that mirror the heights."[11] It is a magnificent reminder of nature's domination over the fragile urbanization spread far below, its skyscrapers puny against the mountain's looming backdrop.

When Curtis first visited the mountain it was already known for something else: its treachery. Plunging temperatures and fierce winds estimated as high as 100 mph sucked warmth from human bodies. Monolithic as Rainier appeared, fractured rock composed much of its skin. Rocks sloughed off, tumbling over trails. Huge boulders crashed down. At times the plunging debris formed a symphony of sorts, "big masses of snow and rock falling thousands of feet at frequent intervals, causing a tremendous roar." Rain, sleet, snowfalls, and sudden squalls made climbers' lives miserable. Climbers had to cross glaciers striated with deep crevasses. Confronted with a crevasse, they had to find a hardened snow bridge over the chasm, a bridge liable to collapse at any moment despite its solid aspect. At times mountaineers had to hack steps in walls of ice too steep to climb.[12] That was not all. If mountain climbing were freed from its perils, making it no more hazardous than a garden party or a kindergarten outing, climbing would not release climbers from their human limits. The dangers of snow blindness and sunburn, and contrarily, frostbite, would still be present. Rarified air would enforce tremendous exertion, often on short rations, producing fatigue and exhaustion. Altitude sickness would induce nausea and vomiting.[13]

How much Curtis understood or cared about any of this is impossible to know. In 1896, though he was not interested in ascending Rainier, he could have perused the two guidebooks to the mountain, its surroundings, and its approaches. Although he was an avid reader and appreciated adventuresome, daring people, his relative isolation at Port Orchard, his hard work there and in Seattle, and his youthfulness probably precluded his learning much about Rainier from mountaineers' accounts. He knew of the first climber fatality on the mountain because his brother Edward led the ascent. Edgar McClure, a professor of chemistry at the University of Oregon, was descending from the summit in the dark when he walked over the edge of a cliff-like rock and fell to his death. The accident was remarkable because it occurred during an expedition of the Mazamas, a Portland mountaineering club noted for its disciplined, quasi-military ascents. A relaxation of the usual discipline during the descent may have led to the accident. McClure's death occurred during the night of July 27, 1897, shortly before Asahel left for the Yukon.[14]

Whatever his knowledge of Rainier, Asahel was in no position to exploit it or to learn more. After his return from the gold fields in 1899 he was a busy young man. Becoming engaged, then married, beginning a family, resuming his commercial photography business, involving himself in photography or photo-related companies in San Francisco and Tacoma, as well as in Seattle, all stole time and energy from enjoying the outdoors. He would not be denied at least one adventure while he lived in San Francisco. In 1903 he joined the San Francisco-based Sierra Club on its expedition to—where else—the Sierra Nevada. More than likely he packed his bulky, fragile camera equipment of the day to record scenes from the heights. Or he may have borrowed or purchased photographs from fellow photographer Edward T. Parsons, who definitely took pictures of the ascent. In any case he returned from the adventure with "very beautiful pictures" of the climb, photographs he shared with a meeting of the Seattle Mountaineers late in 1908. Curtis entertained his audience with a narrative of the beauties and dangers of the Sierra Nevada but it is doubtful that he enjoyed the expedition for its entire duration, which consumed at least three weeks.[15] It is unlikely that a junior employee in the early twentieth century could take almost a month's leave to pursue a hobby.

Curtis was free to climb Rainier once he returned to Seattle and could wangle time from partner William Romans. In July 1905 he joined a group of climbers from the Boston-based Appalachian Mountain Club, the Sierra Club, and Portland's Mazamas, because Seattle did not then have a mountaineering club. Reaching the summit of Rainier was the ultimate adventure of a lengthy excursion that involved much preliminary climbing and camping. The adventurers at last climbed the mountain while they successfully skirted its perils. They spent a brief three hours on the summit, then returned to their base camps without incident.[16]

The episode was far from uneventful for Curtis because he met and formed close friendships with two men who would figure in his future. W. Montelius "Monte" Price was less widely known, and ultimately less important as a friend, though the two would climb Mount Shuksan, close to the border with Canada, the next year. Both men lived in Seattle but apparently were unacquainted until Price, adventuring with the Sierra Club, met Curtis. The common mission promoted group interaction, so it was not unusual for two climbers from different clubs to arrange their blankets beside one another. Curtis and Price talked before drifting off to sleep, a talk fixing Price in Curtis's mind as a potential climbing partner. The meeting with Stephen T. Mather was vastly more significant. Mather in 1905 was 38 years old to Curtis's 31, the scion of a well-to-do California family but well on the way to building a large independent fortune in the mining, processing, and especially the selling, of borax. Households used the refined product as a cleanser but it was, in addition, important in the production of ceramics, glass, and other products. He could not have been more unlike Curtis in appearance. Over six feet tall, with a lithe frame and intense blue eyes set in a handsome face, he was enormously intelligent and furiously driven, a human dynamo. He did not impress people, he bowled them over. His interests ranged widely, from a rah-rah devotion to Sigma Chi, his fraternity at the University of California, to the Art Institute of Chicago and the Sierra Club. He and Curtis bonded, probably becoming "Steve" and "Ace." Mather's friendship was invaluable to Curtis when Mather became an assistant secretary of the Department of the Interior in 1915 and, two years later, the first director of the National Park Service (created 1916).[17]

Curtis and Mather likely met, shared their love of the outdoors, and their conceptions of national park development during a performance staged by the Sierra Club climbers and their Appalachian guests. One evening the "Sierrans and Appalachians, clad in fantastic costumes made up from their camping outfits," raided the Mazama camp. "Borne at the head of the procession was a goat's head, fashioned from the distorted root of a tree, and labeled 'The Original Mazama'" because "mazama" was derived from a Native word for mountain goat. "Great hilarity prevailed, and songs, college yells, and ten-minute speeches by members of the three clubs" punctuated the event. The impromptu theater smacks of a Mather-inspired stunt. Two nights later the Mazamas, dressed as Indians, returned the favor. "On another occasion the Appalachians furnished" the entertainment.[18]

Mather's ebullience had its counterpart in bouts of depression, a cycle known to the ancients but first medically described as the manic-depressive state in the nineteenth century. The complexity of the illness was better understood after Mather's day, with drug therapies introduced beginning in the mid-twentieth century. A later generation of psychotherapists would have diagnosed his emotional oscillations as a form of bipolar disorder, swings between what were originally called the manic and depressive states. Indeed, during the mountaineering high jinks of 1905 Mather was rebounding mentally and emotionally after a "nervous collapse" during 1903-1904. Deeply depressed, he retired to one sanitarium, then another. A later episode, practically coincident with assuming the directorship of the National Park Service, required a third retreat to a sanitarium. Another serious event resulted in his disappearance into his Connecticut estate. In all these instances he was unable to function effectively until reaching a psychological nadir, after which he regained the lost emotional ground in a fairly rapid upswing. It is likely that other, less radical, shifts in mood occurred, possibly passed off as physical illnesses.[19]

Mather's problems concerned Curtis, perhaps equally because of his admiration for Mather and because of his anxiety about the effective administration of the national park system. Probably he considered Mather's lapses to be aberrations in an otherwise brilliant career. In any event he had his wife's disorder to deal with, as well as the pressures of a busy personal and professional life, plus the mountaineering, mountains, and mountain parks that were never far from his thought and action. Mountaineering came into focus during an unplanned meeting with Monte Price in downtown Seattle during the spring of 1906. As Price remembered it, Curtis declared that he wanted to be the first to conquer an unclimbed mountain. The supply of unvanquished mountains was diminishing, he noted, and asked Price for suggestions. Price had none, but as he remembered, Curtis then said, "Well, there's Mount Shuksan. The Mazamas are going to Mount Baker in August. We could go up to the area with them—go off to Shuksan on our own. What say?"[20]

There are significantly differing versions of the Shuksan climb, with a mystery tacked on at the end. When he was "80 plus," Price recalled that only he and Curtis made the assault on Shuksan, he carrying Curtis's big view camera and case, while Curtis lugged his tripod and glass plates. A "sheer wall" confronted them on the first day, so the next day they attempted another route, climbing over rock pyramids, some "so slick" that they hammered hand and footholds in them, instead of the usual mountain climbing chore of using ice axes to hack out stairs in glaciers. At the summit they saw "some lightning-scarred rocks" but no cairns or other evidence of previous climbs. So they built a cairn and returned by sliding down a "chimney," a narrow shaft between two rock faces. Had they known about the chimney earlier, Price lamented, they could have ascended through it instead of scrambling over the difficult rocks.

Price's recollection is dramatic, but Curtis's account, published less than a year after the climb, told a different story. Curtis described Shuksan as "a beautiful mass of igneous rock with cascade glaciers flowing outward on all sides, except the north." The Curtis party, originally five men, ascended from the northwest over a relatively gentle ridge "until we were almost directly under the main pinnacle." Instead of Price's "sheer wall" blocking their path, in Curtis's account "an approaching storm, and the lateness of the hour" forced the party to retreat. "Two days later," not Price's next day, only Curtis and Price worked around to Shuksan's south face. They climbed to about 6,500 feet, spent the night, then reached the summit over what Curtis wrote was "a rather difficult piece of rock-work." The Curtis and Price accounts agree that there was no indication of a prior climb. "We left a record of [our] ascent in a glass jar under the cairn that we built, claiming the ascent in honor of the Sierra and Mazama Clubs of which we were members," Curtis wrote. The vista at the summit should have been "particularly fine," he noted, but smoke from forest fires obscured all but a dim view of Mount Baker's top and the peaks of "a few" other mountains. His brief account mentioned neither the descent nor the often-reproduced photograph showing a bearded Curtis seated—almost sprawled—in front of the cairn, and a mustachioed Price standing nonchalantly, leaning on his alpenstock, to Curtis's right, behind the cairn.[21]

Curtis's narration is closer in time to the climb and must be considered the more authoritative, yet puzzling questions remain. Why did he and Price have to work around Shuksan from the northeast to the south to climb the mountain, if Price's "sheer wall" did not exist? What of the three mountaineers who appear briefly in Curtis's narrative? Curtis identified them as J. H. Lee, Rodney Wilson, and E. G. Grinrod, but they disappeared from Curtis's account after the group's advance was thwarted on the first day. Price never mentioned them. Did the three return to the Mazamas after that frustrating first day? Did they split off from Curtis and Price and search for another route to the pinnacle of Shuksan? Did they discover the chimney that Curtis and Price slid down when leaving the summit? Did they arrive too late to be among the first to claim the peak, well after Curtis and Price had built their cairn and buried their jar? Strangest unanswered question of all, why does the *Seattle Post-Intelligencer* feature include a photograph of Price holding a "Picture of First Conquest of Mount Shuksan" showing five people standing on what appears to be the summit? Why didn't the author of the Price story ask about the obvious discrepancy between his narrative of two climbers who shook hands on a deal to climb Shuksan, and a summit photograph of five mountaineers? Or did he ask, but decide not to complicate a compelling tale of two tough climbers pitted against an unforgiving mountain? It is a minor mystery, one among several enigmas in Curtis's life.

The 1906 climb may have led to the birth of Seattle's mountaineering organization, the Mountaineers, as the climbing club became known. According to Mountaineers' lore, Curtis and Price began their discussion of a club during the climb. The talk spread to others, with Curtis envisioning "a small club of perhaps twenty folks." If the movement were to develop beyond Curtis's modest conception, however, it needed a catalyst. Dr. Frederick A. Cook supplied it. In November 1906 Cook landed in Seattle from Alaska, claiming an audacious, astounding "first," the initial climb of Mount McKinley, accompanied only by a powerfully built, stolid packer. Curtis headed a mountaineer delegation to welcome Cook, who was already known for his brave, if failed attempt on McKinley in 1903, and for polar exploration. Cook attended a November 8 meeting of the provisional Mountaineers Club following, as the minutes read, his "successful climb of Mt. McKinley." Curtis chaired a five-member committee to "communicate with the Mazama and Sierra Clubs" to integrate their interests with the "later" formation of a Seattle organization.[22]

The next night Cook's reception at Seattle's Alaska Club was one "which American audiences seldom give to anyone save a national hero." After introductions from the president of the Mazamas and the mayor of Seattle, Frederick Cook stepped before his audience. Under six feet but giving an impression of height, deep chested and broad shouldered, he possessed a firm jaw and intense eyes. His platform demeanor was not overwhelming for all that, but was calm and self-effacing. Although he diffidently described himself as a novice mountaineer, he claimed an affinity between mountain climbing and polar exploration. The same problems of packing enough food, absence of roads, advancing toward a goal while "nature was in her ugliest mood," a few humans struggling in isolation, made the climbing situation all too familiar. He modestly described his polar inventions adapted to climbing: a superior horsehair rope, an almost weightless silk tent, and lightweight sleeping bags. The slog up McKinley involved incredible hardship. Cook and his companion spent one night on a ledge they hacked from solid ice, one always awake to prevent the other from freezing to death, or rolling off their perch to certain doom. On another night he and his packer "could hear each other's teeth chattering." They husbanded their dwindling supply of food in order to advance toward their goal. Their only disappointment on reaching the summit and surveying the majestic scene spread out below them was the lack of stones with which to build a cairn containing a record of their conquest. It was a thrilling narrative made all the more effective by Cook's praise of the other members of the expedition who turned back while still on the outskirts of the great mountain.[23]

Unfortunately Cook's presentation was mostly hooey, or that is the consensus of those who examined firsthand the route Cook claimed to have taken. Cook could not have reached the summit by that route, they found. Granted Cook's other discoveries in the McKinley area, his inventive turn of mind, his personal courage, and his insights into the problems of human survival in inhospitable environments, the overwhelming opinion of his detractors remains the consensus more than a century later.[24]

Truth had not caught up with Dr. Cook on that emotional night in November 1906. Instead its fervor swept Curtis and other Seattle climbers into a mountaineering club. On November 16 the twenty-two enthusiasts present voted to "form a mountaineering club auxiliary to the Mazamas and the Sierras," Curtis's "small club of perhaps twenty folks." Interest was building, however, while the club underwent a series of name changes indicating its growing sense of independence. A month after the first meeting, with forty-five attending, Curtis read the report of the committee on a constitution and bylaws changing the name to "The Seattle Mountaineer's Club, Auxiliary to the Mazamas." Dropping the Sierra Club did not suggest any conflict with the California organization. The new club recognized Curtis's work and influence by electing him to the board of directors with the largest number of votes. At the first meeting of 1907 Curtis, already known for leading nature walks like those of the Sierrans, was named chairman of the Outing Committee, a large responsibility. On November 15 the club achieved complete independence with its final name change to "The Mountaineers." By the end of the year the membership was 233 and growing, a group large enough to be effective, but also large enough to develop factions, as Curtis would discover.[25]

The Mountaineers nonetheless owed a heavy debt to the Sierra Club and the Mazamas. Like them, it opened its membership to women, unusual at a time when many private organizations barred their doors against women, and nationwide woman suffrage was fourteen years in the future. The Sierrans and the Mazamas precedents were significant because the Mazamas were no lounge club but an organization that exposed its members to the rigors of mountain climbing. Slightly more than half the 151 charter mambers of the Mountaineers were women, including Florence Curtis and Curtis's sister Eva, although the men reserved the important decisions for themselves. Women were supposed to wear distinctive clothing once a climb began, "bloomers," a baggy split skirt covered by another, knee length, skirt. Yet a 1910 photograph shows at least four of the eighteen women clearly visible to be in trousers or bloomers so snugly cut as to be indistinguishable from trousers. All are, however, wearing bonnet hats that identify them as women. To later generations raised on principles of gender equality such distinctions could seem superfluous or even demeaning but in the chivalrous Edwardian era those distinctions marked women as objects of care and concern in dangerous situations.[26]

Both Sierrans and Mazamas held yearly "outings," rigorous mountain climbs, carefully organized and conducted. They sponsored more numerous walks through much less rugged, relatively unspoiled natural areas. As noted, Curtis was leading, informally, similar hikes on Sundays, before a Seattle mountaineering club existed. The Mountaineers continued and expanded his hiking program. In the matter of membership, however, the group diverged from both precedent organizations. The Sierra Club opened membership to anyone who endorsed its goals of outdoor recreation, scientific education about nature, and the preservation of natural wonders. The Mountaineers, though sharing the same purposes, were more restrictive. A proposed member needed the support of two current members. After the candidate's name was published, and no existing member objected, he or she was admitted. The Mazamas, on the other hand, required a prospective member to climb a glaciated mountain and submit proof of the ascent, such as a photograph of the candidate on the summit. The Mountaineers did not make mountaineering mandatory, although the necessary support of two members may have discouraged softies from applying. Beyond the requirements of sponsorship and general approval, membership in the Mountaineers was open, in practice, only to Euro-American men and women.[27]

Meanwhile Curtis plunged into his Outing Committee activities, building trails and organizing outings. He worked hard and effectively, and his work was popular at first. He arranged with the Forest Service to build trails at Mount Baker and Mount Rainier, pleading for funds from counties and private sources, an effort only modestly successful. Despite the financial limitations, he ingratiated himself with Forest Service officials, who were then in charge of the Rainier park. As one wrote to another, "Mr. Curtis and his friends seem to be very reasonable and willing to help on trail work. I think we should show our appreciation by giving them all of the information at our command."[28]

The outings made more demands on Curtis because he supervised their logistics, gathering the necessary supplies, then arranging for their shipment and the movement of the outing's members by rail or water to the closest station or transfer point. Next, and before any goods could move, he contracted for their transport from the station or transfer point to the camps in the foothills and up the mountain. The climbers walked, packing supplies themselves beyond the limits of horse travel. The organization's rules restricted them to thirty or so pounds of personal gear, a heavy burden over rough terrain and at high altitudes. Curtis's leadership skills were already tested and confirmed on climbs on Mount Olympus (1907, see Chapter Nine) and Mount Baker (1908). Once the members of the party arrived at their base camp, Curtis formed them into companies, military style, selected the route of travel, and made all decisions, either by himself in an emergency, or in consultation with his staff. In either case his orders were to be obeyed absolutely and without question. The outings were not for the physically weak or the faint of heart.[29]

Curtis and the committee at first envisioned a three-week exploration trip circling the mountain, beginning at the park's northwest corner.

During two reconnaissance trips, the first running from late August into September 1908, Curtis and his few companions saw that the advancing Carbon Glacier had destroyed a trail that they intended to use. They learned from park rangers that an alternate trail would require a summer to build and cost $1,500 (about $39,000 today). A second trip ended July 4, 1909, with the conclusion that the main party would have to abandon the idea of a circumferential trip and concentrate on reaching the summit from the relatively unexplored north side. The group arranged for a temporary trail across the Carbon Glacier from beautiful Spray Park, the planned beginning point of the outing, to Moraine Park, north and east of Spray.[30]

Curtis was busy organizing the trip, a complex exercise he started at least as early as January 1908. His demanding task began with laying out the general rules for the outing, securing an accurate count of the participants, and determining the volume of food and supplies. Buying those necessities, and arranging with the Great Northern to ship them to Fairfax, an end-of-the-branch-line station about six miles from the park's northwest corner, was not the least of it. He had to contract with a packer to move the supplies from Fairfax to the park's interior. Curtis's calculation of the fee revealed that the trip would not be an adventure for the poor and lowly. He set the expenses for the trip at $40 per person in a year when the average industrial worker's income was $512. Even a century later, in an era of radically expanded leisure, few people could afford either the time or the money to take three weeks off from work. The ninety-two people who signed up for the trip were the elite excursionists of their day. They could have argued that they were not "vacationing" in any sense of the word but "roughing it," testing themselves against an environment often capricious and sometimes cruel. They were, nonetheless, taking a long break from their routines, suspending their earned incomes while spending money to do it.[31]

The outing, already revised to omit the circumferential trip, was changed again. A colleague of Curtis's had a "feeling that there must be something wrong somewhere," and went to Spray Park in advance of the main party. He found the park, intended to be a main camp, "still deep in snow" and unusable. In fact the packer "had killed one horse and crippled others" while bringing supplies to Spray. So Curtis decided to make Moraine Park the base camp, abandoning Spray until outing's end, if then, and directing the shift of the commissary and other provisions to Moraine. On July 17 the group left Seattle, arrived at Fairfax late that morning, and made first camp within the park on a drizzly evening. Curtis sent an advance party to finish the trail over the Carbon Glacier, moved the rest of his group to Moraine Park, and established camp. He then began a series of "try out trips" to view the scenery and "to drill members of the party and try their mettle." Curtis and a few others set up a temporary camp (later Camp Curtis). A "broken shoulder," probably suffered at Moraine Park, prevented Curtis from joining the others while they ascended the summit on July 29 over "the icefield that forms the head of the White [now Emmons] and Winthrop Glaciers." Curtis wrote nothing about his accident at the time, but in a 1937 retrospective another Mountaineer remembered how Curtis was playing baseball when a "lady" ran into him. He fractured his shoulder blade, probably when he fell. His friend Dr. Cora Smith Eaton and another physician were in the party, so both probably attended to his injury. Curtis had to be in agony because he did not participate in what he wrote was the work of the advance party, "the most difficult" ascent "made by members of the club" because the "two men…had to break trail or cut steps so much of the way."[32]

The next day, the advance party's work done, sixty-two hardy or merely brave Mountaineers tried for the summit. This time Curtis would not let his "tightly bound" shoulder stop him as he led the group to a camp closer to the top, although by a different route. His decision is all the more remarkable because he, or someone else, had to lug an added twenty pounds of photographic equipment to the summit. Many years later he confessed to being "practically helpless myself." In the same letter he noted how the warnings of the advance party were so severe that he allowed the climb only because some Mountaineers agreed to assist back to camp anyone who found themselves unable to continue. If Eaton or others objected to his strenuous, even foolhardy, participation, he waved them off. He organized the party into a general staff consisting of himself and five others, and seven companies, each with a captain and four to eight others. Four companies boasted lieutenants, two of them women.[33]

It was a tough climb. A howling wind grew fiercer as the single-file line toiled up inclines increasing to forty-five degrees, stopping to hack steps from compacted snow. Curtis credited his demanding "try out days" with hardening his troops against the elements. Even so, "I passed along the line to see how everyone was taking it." At 13,000 feet the weary climbers escaped from the rising wind in "a half closed crevasse large enough to shelter the entire party." There they rested and ate lunch. Then they trudged east over the top of the Winthrop Glacier to the saddle between Liberty Cap to the north and the East Crater to the south. From there "the ascent was easy," and, once in the crater, "out of the wind," everyone took comfort in the "hot rocks and ashes," pleasant reminders of the once-active volcano.[34]

The next order of business proved frustrating because Curtis had volunteered to plant the official flag of the Alaska-Yukon-Pacific Exposition, Seattle's 1909 world's fair, on the summit. He and some cohorts fastened the flag to its staff in the teeth of the gale. The wind repeatedly blew down the staff and flag, so the climbers took them across the summit to the deep snow of Columbia Crest, where they survived a mere fifteen minutes before the gale snapped the staff. One company remained overnight on the summit. That company took the rescued flag and secured it in the crater, away from the wind. The rest of the party returned to Moraine Park in a two-day descent.

About fifty of the group, Curtis included, ventured to Spray Park. His report praised the moonlit view from the first camp on the descent, as well as the visual glories of Spray, in romantic, well-crafted prose. There was no need for him to discourse on the physical and mental challenges he and his fellow climbers met, the tests of endurance, the pain overcome, and the strengthening of moral fiber weakened through the pursuit of enervating desk-oriented lowland occupations. He did write of two practical concerns that would remain articles of faith with him. The masses had to be granted access to the park, thus "trails and roads" were essential. Yet the masses could not be expected to restrain themselves on seeing Spray Park, "so beautiful that it seemed unreal" Only "the strong hand of the government" could resolve the tension between mass access and mass behavior, and "prevent wanton destruction."[35] For Curtis the greatest benefit was a kind of spiritual self-realization, returning "home the better for the many lessons learned in the solitudes. The trivial things of life, the petty cares that to us seem so great, shrink back in the presence of this majestic mountain. It is as if one heard from out the solitudes a voice: 'Why all this haste? Why all this fret and care?'"[36]

The outing to Mount Rainier was the high point in Curtis's relationship with the Mountaineers, physically and metaphorically. In the midst of his planning for the trip he submitted his resignation as chairman of the Outing Committee, effective at the end of the Rainier climb, to Edmond S. Meany, the club president. He did so with "regret," pleading the need to spend more time with his photography business. His letter suggested combining the posts of secretary and chairman of the Outing Committee, as Cora Smith Eaton would again two years later. His resignation appeared to be prompted by exactly what Curtis wrote that it was, the need to restrict a voluntary, unpaid activity in favor of earning a livelihood. His suggestion about combining the two offices would not have benefitted him personally and would have brought together two important posts. Neither his resignation nor his suggestion alienated any significant number of Mountaineers, as his reelection to the board of directors at the 1909 annual meeting demonstrated. The following July Meany wrote a longhand letter to Curtis transmitting the Mountaineers' "high appreciation" of his "magnificent leadership during the Outing of 1909." Meany included a "token of our regard," a "binocular field-glass which we hope may be your companion on many a joyful climb to 'The Great White Hills of God.'"[37]

But there would be no more climbs with the Mountaineers. A fierce public dispute of 1910 over his timely presentation of the expense reports of past climbs, including most recently the Rainier ascent, led to charges from some other board members that his accounting methods were not businesslike. There was no suggestion of fraud, but rather concern about unconscionable delays. Curtis's friend and fervent supporter Cora Smith Eaton defended him as the club's best manager, who had to grapple with the complexities of the Mountaineers bookkeeping systems as well as the almost overwhelming demands of his own business. Curtis confirmed that his business had grown "so large" that he could not devote as much time and effort to the club as he had in the past. Although he helped with the preliminary planning, he did not join the July–August 1910 ascent of Glacier Peak in the Cascade Mountains, perhaps to steal precious hours to square his accounts with the Mountaineers.[38] He did not participate in later climbs. He failed to be reelected to the board of directors at the November 1910 annual meeting, where he presented himself as a candidate, possibly because of lingering resentment about his delayed accounting, but more likely because he was unequivocally on record as unable to assume his old responsibilities. He did not lose because he wanted the club to endorse reduced boundaries for the Olympic National Monument on the Olympic Peninsula, a reserve created by President Theodore Roosevelt in 1909 (see Chapter Nine). The argument over reducing the boundary did not emerge until 1912.[39]

It is possible that Curtis tried to succeed Meany as president and was rebuked. According to one Curtis biographer, "whatever he joined, he usually tried to run," but it is unlikely that he would have attempted to unseat Meany, a friend to many and a friend to Curtis until Meany's death in 1935.[40] What is more likely is that a majority of the Mountaineers noted his declining involvement with the organization, his interest in an extensive road and trail program for Rainier, and his lack of enthusiasm for an Olympic National Monument closed to development.

Neither Curtis nor any dominant group behaved as though he were ostracized from the Mountaineers. In April 1911 Eaton wrote to Meany concerning "Mr. Curtis and his loyal services…It is a fact that every program offered so far by the Program Committee has been suggested by Curtis to me and I have passed it on," including future programs "as far ahead as there are any plans at all." Eaton reported how Curtis had secured trail information from a man about to leave Seattle, after attempts by another member failed. Eaton maintained that there "is no man in the club but Curtis who can adequately maintain your ideals." She spoke for "several" members who worried about the Mountaineers being "weak in the two important offices, Secretary and Chairman of the Outing Committee." Eaton's solution combined the two under Curtis as the Sierra Club organized them under William E. Colby, paying Curtis enough salary and expenses to make the job attractive. Eaton admitted that "there might be some defections" if Curtis took on the task but "we could well afford to lose these well-meaning but narrow-minded members rather than have the pattern of our club cut down to their standard." Curtis might not take the job, "but in this direction lies the hope of our club." "Without the proper support" that Curtis would supply, Meany would not wish to continue in the presidency.[41]

Meany's reply revealed why he became president the year after the founding of the Mountaineers and remained president until his death. He assured Eaton of his sympathy with "most" of her letter. Yet he noted how he had "helped to steer our Club along the lines of least resistance," resulting in an "absence of contention." He "would have to run away" from a contentious club because his "university work" would not "permit weakness through other cares."[42] Meany would do nothing to jeopardize his position above factional divisions.

Curtis faced a similar problem in his continuing effort to move Meany toward involving the Mountaineers in developing roads and trails in Rainier. Curtis told Meany that in promoting the park, the club could "do a great work," because "it will always be hard" to persuade Seattle and Tacoma "to unite upon any project."[43] Meany encouraged Curtis to write to Secretary of the Interior Walter L. Fisher regarding improvements to Rainier. Curtis did so, stressing the need to make Rainier accessible to the people, who would then realize its value. He also plumped for a "Bureau of National Parks" and a "National Park Commission." Fisher replied that most of what Curtis suggested would require money, and that Congress would not appropriate the needed funds without demands from constituents. Fisher's letter spurred Curtis to a flurry of activity. He wrote the chairman of the Progress and Prosperity Committee of the Seattle Chamber of Commerce about the need to pressure Congress. A *Post-Intelligencer* article carried his exhortation to the people of Washington to begin a propaganda campaign for the park. Meany received his most urgent request, that the Mountaineers "should start a campaign" to persuade Congress to appropriate enough money for road surveys, at least. Curtis urged a special committee, naming another Mountaineer to chair it, "to carry on a steady campaign such as the Sierra Club has for years." He told Meany that members of eastern mountain clubs would write their congressmen "to aid in the work if they received a request" from the Mountaineers. Curtis closed his letter by asking Meany to "bring this matter before the club at an early date," but if Meany initiated a campaign, there is no record of it.[44] As Meany's biographer wrote, "he never became a vociferous or aggressive advocate for national parks or conservation measures."[45]

Curtis made another attempt to move the Mountaineers. His article, "The Future of the Rainier National Park," appeared in the 1911 issue of *The Mountaineer*, together with one of his spectacular photographs of Rainier and its reflection in a lake. His article repeated the arguments for a concerted publicity campaign in his letter to Meany, now addressed to the membership at large. "The most pressing need in mountain affairs in Washington at the present time," he began, "is the improvement and extension of the roads of the Mt. Rainier National Park." He explained that the best way to make the park accessible and develop the necessary nationwide support for it was to provide roads. He urged the Mountaineers to "lead" with "vim" rather than "leave important work for the commercial bodies" of Seattle and Tacoma, because the problem was national, not regional. He asked the club to formulate a "plan of action that can be followed for years," one including roads safe for car travel. He concluded with a nod to "the true Mountaineer" who "would much rather see the mountains from the trail or the unexplored wilderness." The need, however, was "to make the mountains...popular, to get the majority of people into them," therefore it was "necessary to have roads."[46] There is no record of any concerted action based on Curtis's article.

Curtis strongly suggested that elitist mountaineering and strict preservationist sentiment were inadequate bases for popular support of the national parks movement. It's probable that a majority or a substantial minority of club members agreed with him. Had there been a strong reaction in 1911 against Curtis and his democratizing impulse through expanded car use, it is doubtful whether the article would have appeared. *The Mountaineer* celebrated the achievements and the potential of the Mountaineers, but it was not a medium for airing their differences. Meany's moderation and disinclination to promote discord usually muted factionalism, thus it is doubtful if Curtis's exhortation represented an extreme minority view at the time.

Yet blaming Meany for not moving the Mountaineers into activist territory would be too simple. The club was, for Meany, an essential escape from the burdens of the university work he loved, but work producing its own anxieties. In the Mountaineers he cut an impressive figure, at six feet, three inches towering over practically everyone of his era. He was generous with his time and money. Twelve years older than Curtis, he was strikingly handsome in his young manhood, retaining his slender, athletic frame well into middle age. In later years he sported a graying Van Dyke beard. Some photographs of him rusticating with the Mountaineers reveal a sort of urbanized John Muir, the famed naturalist and head of the Sierra Club. By 1911 Meany was a fixture at his beloved University of Washington and a noted regional historian as well as a good bass singer, an excellent raconteur, and a prolific dedicatory and commemorative versifier in iambic pentameter. Understandably, he preferred to advance the club's immediate interests while maneuvering it away from factionalism and controversy.[47]

Curtis was soon relieved of any disappointment with Meany when an initiative from Tacoma overturned his conviction that Seattle and Tacoma could not work together for the improvement of Rainier. As developed in the next chapter, the group, chaired by Curtis, played a significant role in the park's improvement. Meanwhile, Curtis joined a movement to redefine the Olympic National Monument, an action leading to his leaving the Mountaineers, although the outcome could not have been foreseen at the time. The movement began with the Seattle Chamber of Commerce and other commercial organizations campaigning to reduce the size of the Olympic National Monument, or open it to exploitation, or both. As things stood, only those mining claims entered before the monument was created were legitimate under the law and administrative interpretation. The goals of the organizations were, first, to return a large portion of the monument to the Olympic National Forest, out of which it was carved, and second, to allow mineral development within the remainder of the monument.[48]

Curtis's role in this activity cannot be understood apart from his beliefs as they had matured by late 1911 and early 1912. First, he believed in conservation, that is, conserving natural resources for their rational exploitation under federal supervision and control. His belief in conservation for ultimate use meant that he would champion the development of Washington's timber and mineral resources. Development would spur industrial expansion, the concomitant growth of business linked to timbering, mining, and industry, and, therefore, an increase in population and prosperity in the state. Second, he believed in preserving scenic wonders, mountains, foothills, forests, and meadows of inestimable value, and animals threatened with extinction or with a reduction in numbers so radical as to make them practically invisible to the humans sharing their habitat. Saving animals promoted tourism and related economic activity such as lodging, road building, and vehicle maintenance. There were psychic advantages, too. His daughter Betty recalled how her father "found his beliefs, and his religion in God's world and the forests and mountains. He always believed that if man could just stay close to nature he would survive." His convictions about the uplifting and sustaining powers of natural scenes derived from generations of nature writers who promoted a retreat to uninhabited regions as the antithesis of urban life.[49]

Curtis's beliefs in controlled exploitation, preservation, and opening national parks to the masses at times conflicted with one another, forcing him to choose or to compromise in specific, on-the-ground situations, and opening him to criticism from absolutists of all stripes. In the case of the Olympics, he believed that no national park in the mountains was possible without some compromise between preservation and exploitation. The alternative was a negative situation, no development of potentially valuable resources and no conversion of the monument into a permanent national park. Curtis's beliefs relate to the struggle to return much of the Olympic monument to the national forest, where its timber would be available for harvesting, meanwhile opening the reduced monument to mining and its accessory activities.

Late in 1911 the movement coalesced into a group composed of ten commercial organizations, including the Seattle Chamber of Commerce and the Tacoma Commercial Club. Curtis persuaded the Mountaineers to join the group and, at Meany's request, served as the representative of the club. He was also the delegate of the Seattle Chamber of Commerce. Congressman William E. Humphrey, a Seattle representative who was interested in a national park, asked three men, including Curtis in his dual role, to thresh out a program acceptable to all. The program, necessarily, would have to include guarantees for some commercialization of the Olympics.[50]

Curtis believed that he faced several unalterable conditions in his attempt to wrest a national park from the monument. Congress would not pass an Olympic national park bill over the opposition of the natural resource developers. Congressman Humphrey had tried that, introducing bills to create a game preserve in the Olympics. The bills failed. Humphrey next successfully appealed to President Theodore Roosevelt to establish the Olympic National Monument partly as a game preserve. Roosevelt created the monument by proclamation on March 2, 1909, one of his last days in office. The monument was inadequate to protect the "Roosevelt" elk, a subspecies, because Congress would not approve converting the monument into a national park large enough to include the entire elk range. Besides, a 1905 Washington law prevented the killing of

Roosevelt elk, so a vast wildlife preserve was, in theory, neither feasible nor necessary. Finally, there was no hope for a park bill unless, within the proposed park, "the mining interests were protected." Curtis did hope to enlarge the park to the east, "to have the high peaks seen from [Puget] Sound" made "a part of the park," and, incidentally, preserving a large area of timber. He and "quite a number of individuals" wanted the eastward extension and it would have happened "had it not been for the opposition of people who have [timber] interests in that section."[51]

Curtis and other Mountaineers had worked with Humphrey on his proposed legislation, "a step toward a national park"; therefore he was at first enthusiastic about the national monument. It encompassed what he believed to be 642,000 acres carved from the Olympic National Forest. (It was, in fact, approximately 610,560 acres.)[52] On March 4, two days after Roosevelt's proclamation, Curtis wrote a fellow Mountaineer "that our work in behalf of the Olympic National Park is over, and that the park is an assured fact." Curtis's optimism led him to overlook the reality of the monument, which was not a park at all but a huge tract in the hands of the Forest Service, an agency caring little, at the time, for the wildlife and scenic preservation impulses behind its creation. Curtis was more concerned with the size of the monument, a forecast of his future antagonism toward a large park. "The park," he wrote, "is much larger than the one we had planned, some 80,000 acres larger."[53]

Curtis would attack the monument soon, principally for two reasons. First, its stated purpose was to protect the magnificent Olympic or Roosevelt elk, a variety larger than the typical elk, but the monument could not include all of the elk's range without impinging on commercially valuable timber. He had no wish to see the elk slaughtered or deprived of forage, but if the issue came down to animal or human benefit, he would opt for the people's access to a useful resource. Second and more important, the monument closed potentially valuable mineral areas to practically all exploitation. Some way had to be found to develop a park smaller than the existing monument while opening it to mining. As noted, Curtis was one of a committee of three that began in 1911 to draft a proposal for converting part of the monument into a national park, but with a provision for mineral extraction.[54]

Curtis's only hope for a national park lay in a compromise acceptable to the commercial organizations and to the Mountaineers. He went first to the "commercial bodies...opposed to any National Park" and persuaded them to accept a park about half the size of the monument. Converting the exploitation-minded required all of his persuasive oratory. "I went to a meeting where no one would consider any discussion of a National Park and came away with a signed request for one." The deal included what the "commercial bodies" did not want, a national park, and what they insisted upon, placer and lode mining under permits allowing for related timber cutting, road and rail transportation, and dam construction, or as it was then called, water power. The report detailed the various compromises involved with excluding, or in one case, including land in what worked out to be a proposed 334,000 acre park. It asserted that "discoveries of mineral wealth of great value have been found at many points" in the Olympics as justification for the mining. Mining roads would help overcome the lack of roads and trails in the monument, while the park's hunting restrictions would protect wild animals. Moreover, the area's "unusual scenic beauty" made it "of more value as a national park than for any other purpose." It would "attract visitors that will be a source of revenue to the peninsula and the state."[55]

Curtis's greatest coup was bringing the Mountaineers on board, or so he thought. It is obvious from Meany's statement in the 1911 *The Mountaineer* that the club was involved in the negotiations. The Mountaineers, Meany wrote, "felt secure in the fruits of their labors," in helping to gain the Olympic monument "until there arose complaints that prospectors and miners were hindered in their efforts to secure the mineral wealth supposed to exist within the Monument." To resolve the issue, the Mountaineers and the commercial organizations "amicably agreed" to "permit mining and prospecting under proper regulations within" a new Olympic national park. They agreed to limit the park's size. The park would "include the summits of the mountains and as little as possible of lands useful for agriculture and forestry." Meany related how the people "interested" in mineral development were surprised and grateful "that the Mountaineers took such liberal ground," but "such has always been the attitude of the Mountaineers." Meany concluded with an oblique reference to the "thousands" who would visit the Olympic park, and could, in the future, prevent any development bringing "harm" to its beauties.[56]

Then the Mountaineers' board of directors reneged on the deal. Curtis was shocked and outraged when the arrangement he had so carefully crafted fell through. This is what appears to have happened: the board appointed a committee to review the report that Curtis negotiated, a majority of the committee strenuously advised the entire board to reject the report, and the board went along. Among the reasons for rejecting the report may have been concerns about the compatibility of park preservation and mineral exploitation, or, specifically, the dams and stream impoundments essential to realize the "water power" concessions to the mines, or the exclusion of scenic territory worthy of inclusion in the park, or all of them. As one of the officers in the Sierra Club wrote to Curtis, the concessions to exploitation were "practically the same as throwing down the bars and having no park at all." Further, the "limitations and exceptions" would "establish a most vicious precedent with relation to all the other national parks."[57] The statement is overwrought, for Congress admitted Glacier National Park in 1910 without any such precedent, but with practically identical provisions for future resource development, provisions resulting in a dam despite Mather's opposition.[58]

Curtis saw another motive behind the board's action. In a furious letter to a member of the review committee, L. A. Nelson, he claimed that some "members of your committee were influenced by their hatred for me." In a statement reflecting Eaton's earlier letter to Meany, Curtis charged that "two or three men" were dictating to the club and running "the whole show. Give them an invitation to get out." As for himself, Curtis concluded, "I will never be back in the organization again."[59] Nelson attempted to placate his understandably angry friend. The board's refusal to adopt the report, Nelson maintained, came from two sources, a concern to study the report carefully, and personal antagonism toward a member of the Curtis park report committee, who "should have been friendly" to the Mountaineers, but was not. There was no expression of dislike for Curtis, indeed the Mountaineers' committee told the board "that you had done everything you could to help us."[60]

Curtis would not be mollified. It fell to him to tell his committee that the Mountaineers had repudiated Meany's position. His news provoked "a stormy hour or two when I thought the whole deal was off." The first counterproposal was to ask the commercial organizations to reject the national park and send a "resolution to Congress asking for the abolition" of the monument. Curtis saved the day for a park with the mining concessions because he was convinced that the only hope for a park was to keep mining and prospecting within its boundaries. He wrote another correspondent about his continuing belief that the Mountaineers were "influenced in a measure by a number of the members who have taken a strong dislike to me." True to his word, Curtis did not return to the Mountaineers. As late as 1915 he was working to have his compromise measure enacted while opposing the club's alternative, a park bill shorn of development provisions. Neither became law, and no Olympic park measure would succeed for many years (see Chapter Ten).[61]

Curtis's relationships with the Mountaineers and the Mount Rainier National Park formed a significant part of his life. He was devoted to Mount Rainier before the park existed, and was an avid mountain climber. Mountaineering provoked one of his few reflective statements on any subject. He refused to let a "tightly bound" broken shoulder prevent him from reaching the summit of Rainier during the group's 1909 climb. His indifference to pain when it would have been easy to pass the leadership of the ascent to another Mountaineer demonstrated his courage and determination. His earlier, active role in the founding of the Mountaineers spoke volumes about his commitment to preserving natural scenery. "Many Mountaineers seem to forget or maybe they never knew that Asahel Curtis really started the club," one member wrote almost fifty years later. Others were involved in the founding, but Curtis played a major role even as increasing enthusiasm outran his vision of a small club.[62]

The circumstances surrounding Curtis's retirement from the Mountaineers force a reassessment of some prevailing accounts of the break between the man and the organization he helped to found. One matter involves the problem of national park development. Perhaps Curtis and a majority of the Mountaineers were diverging in their views of the proper way to improve the Rainier park or any national park, but to believe that they were headed for a complete break on the issue is to believe in the inevitability of events. Curtis argued for roads in Rainier if the park were to serve the needs of the typical recreationist. The Mountaineers, or some of them, later defended a limited-access park designed principally for the hiker and climber. These positions were not, however, mutually exclusive in 1912.

Concerning the unrealized Mount Olympus Park, Curtis's letters refer to a lively discussion within the Mountaineers about the board of directors' action.[63] The board rejected, not just his hard-won compromise, but its president's commitment to that compromise. Had there been a consensus on preservation within the club there would have been little or no discussion. Whatever the dialogue, and whatever soothing words L. A. Nelson or others wrote or said to Curtis, the board's decision stood, indicating a preservationist majority, or if not that, a majority willing to accept the demands of a minority of preservationists. The club did not have the last word in 1912, however. Preservationist sentiment within the Mountaineers and preservationism in the larger world were of different degrees, as the group discovered when they proposed an Olympic park bill deleting the mining and prospecting compromise. Its failure demonstrated the limits of the preservationist approach, compounded by what could be construed as the Mountaineer's deliberate insult to the commercial organizations with which it was once allied. One member of Curtis's Olympic park committee wrote of how the club was "very largely made up of school mam's [sic] and people of the fanatical conservation type, who would have been pleased to set aside the whole middle west for the benefit of the buffalo."[64] The statement was a caricature, but probably reflected regional business attitudes in the wake of the Mountaineers' recanting of an apparently assured arrangement.

Lastly, statements suggesting that the Mountaineers as a group became alienated from Curtis are not responsible assessments of what transpired. Assertions that Curtis lost "the confidence of the membership" of the club, or that his committee's recommendations were dismissed "out of hand," or that "in part he was driven from the Mountaineers" misrepresent the situation.[65] He could have stayed. Meany, who surely was acutely embarrassed over the board's abrogation of his understanding with the development groups, remained president of the Mountaineers for another twenty-three years. But it was not Curtis's way to suffer in silence or abjectly accept what he believed was a deliberate personal humiliation. Once out of the organization he did not carry a grudge, but instead cooperated with the Mountaineers when their interests converged. He would do so, however, as president of another organization that would become the Mount Rainier National Park Advisory Board.

Chapter Seven

Mount Rainier: Founding and Leading the Rainier National Park Advisory Board

Curtis plunged into Rainier events ever more deeply when he assumed a leading role in founding the organization that became the Rainier National Park Advisory Board (RNPAB), which he led for sixteen years. His chairmanship centered on Rainier's road and trail plans, no simple matter because roads and trails involved far more than engineering issues, but instead concerned basic questions of the park's nature. Would Rainier be given over to tourist cars, or would autos be restrained in favor of preserving the most scenic areas for campers, hikers, and mountaineers? There were no easy answers. Once agreed upon, roads and trails had to be financed and constructed, bringing fresh questions of administrative and Congressional politics to the fore. Roads and trails related to every other development within the park, including the activities of the concessionaires, their transportation system, their hotels, and their campsites. Road location entangled Curtis and the RNPAB in the usually difficult and sometimes maddening matter of connecting park roads to access highways outside. Controversial as some of these issues were, the contentions usually involved intramural struggles among people who sought the same or similar ends.

The future RNPAB gestated in 1911 when the rival cities, Seattle and Tacoma, united to promote the park. Their chambers of commerce joined to create an organization designed to quell their competition and unite on one issue, developing Rainier. Together, they determined to take advantage of the scenic masterpiece on their doorsteps, an opportunity for two cities uniquely close to a major national park. Their larger rivalry dated from the 1880s and involved railroad connections as well as water transportation on Puget Sound. Population figures told the story of Seattle's triumph. In 1890 the two were of comparable size, with Seattle slightly larger. By 1900 Seattle's population increased to 80,671, versus Tacoma's 37,714, largely because Seattle was the most developed American port nearest the burgeoning gold fields of Alaska and Canada's Yukon. By 1910 the contest was over. Seattle tallied 237,190 to Tacoma's 88,749, a margin that increased in subsequent decades.[1]

The first initiative came from Tacoma, the defeated antagonist, in December 1911. Until then Curtis took a dim view of Tacomans who, he thought, wished to control access to the park through Tacoma. An experience earlier that year confirmed his belief. In September 1911 he wrote a letter to Walter L. Fisher, the Secretary of the Interior, at Fisher's request, outlining needed improvements to the park. Among other betterments, Curtis advocated a road along the Carbon River at Rainier's northwest corner. The road, if connected to passable roads in the park, in the surrounding national forest, and from Seattle, would allow Seattleites to reach Rainier without passing through Tacoma. Curtis's object was to improve access to the park, not to spite Tacoma, because he also put forward widening and extending the park road from the rival city. When he "tried to get the co-operation of Tacoma" for the Fisher letter, however, he was rebuffed. "They thought that I was going to pull off a road to Carbon [River] and they were afraid of it." This experience impelled Curtis to urge Edmond Meany of the Mountaineers to take the lead in Rainier improvements with the statement "It will always be hard to get the two cities to unite upon any project."[2]

The breakthrough came from Thomas H. Martin, the manager of the Tacoma Commercial Club and Chamber of Commerce. On December 23, Seattle's morning *Post-Intelligencer* published a letter from Martin, who pressed for a meeting between the commercial organizations of Seattle and Tacoma. Such a meeting, he declared, "would result in a businessmen's agreement to work together for a full and complete development of our great national park." Chiding the *P-I* for an editorial claiming that "certain common Tacoma interests" were opposed to a northwest (Carbon River) entrance, Martin asserted that the *P-I* was mistaken, that the opposition wasn't representative of either city. Action on Martin's proposal quickly followed with Curtis an activist.[3]

Curtis was not then entirely disillusioned with the Mountaineers but found Martin's explicit call for Rainier's development a challenging alternative to Meany's temporizing. On March 7, 1912, a joint committee of chambers of commerce and Rotary clubs recommended an organization with "permanent machinery" designed to advance the park improvement agenda. By that time Curtis had broken with the Mountaineers and was fully committed to the businessmen's approach. A week later the organization matured during a meeting at Seattle's New Washington Hotel. Its program included a call for a bureau of national parks, a well-worn, nationwide request. Another item asked for "a complete system of roads" in Rainier including road improvements on the south and west sides of the park. Trails were incorporated in a call for roads and trails enabling rangers and tourists to travel along the uplands, enjoying all the mountain meadows and other sites without returning to the lowlands. It was an ambitious program later substantially realized in the circumferential Wonderland Trail, though never in roads. Finally, the group backed a timber exchange policy; private owners of merchantable trees along public highways swapping their harvesting rights for comparable stands in the national forest. The policy would preserve roadside trees for the enjoyment of travelers. All these proposals and others were not in themselves new. Engineer Eugene V. Ricksecker proposed a circular drive during Rainer's early years, while Mather urged scenic preservation along roads leading to the national park.[4]

There were parallels between Curtis's work in the businessmen's organization and the Mountaineers. Both groups centered on Mount Rainier National Park. Curtis was originally skeptical of the Mountaineers becoming a large organization just as he was at first doubtful of any agreement between Tacoma and Seattle. Once convinced of their permanence and expansion he assumed leading roles in each. Both groups struggled with name changes before settling on those best expressing their purpose. Disagreements over the nature of an Olympic national park caused Curtis's exit from the Mountaineers. His support of a later Olympic park, different from the reserve that the National Park Service desired, alienated him from the service and its parent, the Department of the Interior (see Chapter Ten).

At first the business committee members had some difficulty agreeing on the name for their organization, or knowing exactly what the name was, whether "Joint Committee of Seattle-Tacoma Rainier National Park Policy," "Seattle-Tacoma Mountain Policy Committee," or "Rainier

National Park Development Committee." By April 25, 1912, it was the "Seattle-Tacoma Rainier National Park Committee," with Curtis as the chairman of its executive committee.[5] The committee's use of "Rainier National Park" then and later represented a concession from Tacomans who pushed for banishing "Rainier" from the mountain and the park and substituting the name "Tacoma." Sentiment in Seattle, as well as it could be gauged, favored retaining "Rainier." The simmering dispute between Tacoma and Seattle did not depend on the cities' relative size or prosperity, for Seattle's economic dominance cut little figure in the naming rights to the mountain and park. The conflict flared the 1920s with disastrous, if temporary, effect on the businessmen's cooperative effort to develop the park. It will be considered later. For the time being the name "Rainier" became a neutral identifier allowing both cities to claim the park as part of their hinterland, as a psychic, visual, and tourism dependency, rather than an attached natural resource domain.

At a December 1912 meeting the group asked the Tacoma Commercial Club to join them, which it did. The organization then changed its name to the Inter-City Committee, an unsatisfactory designation because it omitted Rainier, the object of its activity. Early in 1919 Martin and Curtis proposed an expanded organization composed of commercial bodies in the communities surrounding the park, to be named the "Rainier National Park Advisory Board." By March 19, 1920, the two had succeeded in creating the new board with Curtis the temporary chairman, although he did not begin using the title "chairman" until later.[6] The board's immediate business was to arrange for a company "to take over concessions in the National Park," a goal achieved by the time Curtis wrote those words in November 1920. Anti-monopoly challenges to the Rainier National Park Company (RNPC, founded 1916), continued for a few years afterward. The board's long range goals were to increase appropriations for the park, and, as the new name suggested, to offer recommendations on all subjects regarding Rainier. The two goals, increasing appropriations and offering advice, overlapped and were not themselves new. To be achievable they required close cooperation with other organizations interested in the park. Harmony was a requisite. As Martin wrote Secretary of the Interior Franklin K. Lane in 1913, the function of the (then) Inter-City Park Committee was advisory. It did not exist "to make demands."[7] Close cooperation between the advisory group and all other interested parties remained the norm, despite occasional discord, from its founding in 1911 until Curtis formally disbanded it in 1937.

The spirit and the reality of cooperation were remarkable, given the range and variety of the affiliated or related entities. The board itself included fourteen groups by 1933, mostly though not entirely commercial organizations. Any proposal or program had to be threshed out within the RNPAB before its presentation to one or more of the welter of interested associations. Curtis later made much over the board's harmonizing divergent viewpoints, then offering a unified presentation to, for instance, the Park Service. He contrasted this approach with each constituent group's advancing its own agenda, resulting in a series of uncoordinated propositions. Curtis's contention was compelling if not always accepted by others. Yet agreement was the dominant note. There seems to have been little discord among the advisory board members at their infrequent meetings. Moreover, Curtis presumed to speak and act for the board in the everyday application of agreed-upon policy, or in a few instances, in the absence of a policy.[8]

To discharge his duties as chairman of the advisory board, Curtis entered a thicket so tangled, deep, and dense, he should have likened it to a swamp. Some of the board's difficulties arose because it concerned itself primarily with roads, and roads related to almost everything else in the park. Worse for the equanimity of Curtis and the board, road connections outside the park were a responsibility divided between the commissioners of Pierce County, the Forest Service, and the state of Washington. State road development depended, in turn, on agreement among the legislative branch and the governor, an agreement not always forthcoming. Especially galling was Governor Roland Hartley's refusal to authorize a state highway to the park's northwest corner until his inaction and other circumstances doomed the linked road within the park. From 1916 to 1925 the intra-park roads were entirely the responsibility of the Park Service. When the service assigned road construction but not planning to the federal Bureau of Public Roads (BPR), it created another bureaucratic layer.

There were plenty of other federal layers, bureaucratic and personal, for Curtis to cope with. The park superintendent represented the first level of officialdom. The office was something of a revolving door until the supremely competent Owen A. Tomlinson became superintendent in 1924, remaining in the post after the demise of the RNPAB. Curtis often worked with the dynamic director of the Park Service, Stephen T. Mather, in a generally compatible way. Relationships with Mather's successors were less harmonious. The Secretary of the Interior had many responsibilities beyond the national parks, but that official was closely involved at times with Rainier and the RNPAB. Congress controlled the Park Service purse, but the man most closely associated with that responsibility for most of Curtis's tenure was Louis C. Cramton, chairman of the subcommittee on the Interior Department of the House Appropriations Committee. Cramton was the "go to" man in Congress, for without his approval no national park funding request survived. A Michigan Republican from a rural area east of Lansing, Cramton was a friend of the national parks and a fair-minded man. Yet the combined weight of the Park Service and friends in Congress was not enough. The RNPAB and its predecessors needed a full-time lobbyist in Washington, finally finding one in the person of J. J. "Jack" Underwood (see Chapter Eight).

To return to regional relationships, Curtis in his chairman's role collaborated most closely with the general manager of the RNPC, Thomas H. Martin. Martin's company was an anomaly in the RNPAB because it was the Park Service concessionaire, a connection unique in the RNPAB's membership. The 1920 reorganization of the RNPAB's predecessor brought the company into the advisory board, giving it a voice in policy at the same time that it pursued its own interests according to its contract with the Park Service. The connection began in 1915 when Martin, Mather, and a group of businessmen from Tacoma and Seattle agreed to organize a monopoly company to control accommodations, sales, and group travel by bus, rail, or truck inside the park. The RNPC was incorporated in time for the 1916 season. Martin left the Tacoma Commercial Club and Chamber of Commerce to become the company's general manager.[9]

Mather's policy of granting no-bid private monopolies in the national parks has been criticized, but in the case of Rainier it was justified, given the dependent, fragile nature of the company. The RNPC was chronically undercapitalized. Small contributions and park advertising aside, it was unsupported by any railroad. The Great Northern owned a branch line to the hamlet of Fairfax north of the park's northwest corner, while a branch of the Milwaukee Road ended at Ashford, a town west of the southwest corner. Neither railroad showed any interest in extending its line nearer the park or building hotels unlike, for example, the Great Northern in Montana's Glacier National Park. The RNPC's principals were wealthy by local standards but incapable of the huge expenditures essential for a hotel or chalet system. The company required the continued cooperation of interested Seattle-Tacoma business people despite the cities' commercial competition and quarrels over the name of the mountain and its park. It needed the Park Service to smooth its path by refusing to renew the contracts for the pre-existing tourist hotels and camps, allowing the RNPC to buy them at advantageous prices. It needed the Park Service to enforce its monopoly on transportation, as when park rangers required all other hire vehicles to unload and transfer their freight and passengers to company carriers. Private cars, first admitted to the park in 1908, escaped the ban on multi-passenger transportation, much to the company's detriment.[10]

The year of 1924 saw 161,473 people visit the park, all but 12,494 arriving by car, and all save 33,138 from Washington State. Of the 148,971 who drove cars or rode in them, 86 percent were from the state (57,055 from Seattle, 32,474 from Tacoma, and 38,806 from other places in Washington). By then the Park Service had established public campgrounds to supplement the hotels and camps of the RNPC, and "fully 90% of the park visitors desired to camp in the public camp grounds."[11]

Historian Theodore Catton found most visitors who came by car shunned the RNPC's offerings, preferring the free and improving campgrounds of the Park Service. The era of roadside auto camping was ominous for the company, even during its good years. In the autumn of 1924 Martin went by car from Yakima, east of the park, to points high up in the Cascades. The road to Rainier's east side was seven years in the future, yet "up the eastern slope of the Range I saw the evidences of summer camps. Literally thousands of people had traveled up the mountain side" to escape the summer heat of eastern Washington, "and there camped." The mountain was not the then inaccessible Mount Rainier, but a lesser peak to its east. Martin thought that the widespread individual camping boded well for the park. It did, but not for high-end resorts like the company's Paradise Inn. Indeed, the Park Service's and the company's camps were a response to the individualistic disorganization that Martin saw, with its bad sanitation and disturbance of natural areas. Such scatterings of trash, used-up tires, other auto detritus, human waste, and garbage could not be tolerated in Rainier. Once in the park, automobilists had to be funneled into regulated campgrounds. Despite that, and despite the short summer season and the slow growth of winter sports, the company did well enough with its Paradise Inn and other accommodations until the Great Depression forced more visitors into cheaper quarters.[12]

Curtis's relations with the company were, at first, cooperative but not very congenial. He found the accommodations and management in the park seriously lacking both before and after the RNPC acquired them. Early on he criticized the National Park Inn at Longmire, a few miles inside the park's southwest corner, and the Camp of the Clouds at Paradise, a jumping-off place for climbs to the summit. The Camp of the Clouds was known informally as "Reese's camp" after John L. Reese, its proprietor. Curtis's complaint about the National Park Inn was that its management tried to keep tourists out of the scenic Reese's camp, where the camp's guides hustled climbers to the summit in one day from the Reese camp, a climb "possible for the average tourist to do" but not "without difficulty." His critique of Reese's camp focused on sanitation. Flies in the dining tent, "a big mud puddle full of filth" near the tent, and numerous complaints about cleanliness in the camp "would lead me to believe that little if any effort was made to keep it in a sanitary condition."[13]

If Curtis expected the RNPC to improve the situation right away he was disappointed. At the end of the company's first full season (1917) he wrote Mather that "the new company failed to accommodate the tourists." Because he was head of the RNPC's guide service in 1917 he admitted that he was "too close to get the right perspective" although the "multitude of complaints" suggested that inexperienced company workers were overwhelmed. Curtis thought much better of his own work. He reported to Horace M. Albright, Mather's assistant, that he "made fifteen trips to the Summit, taking up about ninety-five people, and more than a thousand through the Glaciers," all "without the slightest accident." The season's one fatality occurred when inexperienced city people not connected with Curtis or any other guide attempted to cross a glacier, and a young woman fell into a crevasse.[14]

Curtis applied for the job of park superintendent at the end of 1917, citing his belief that the present superintendent planned to resign and enter military service. Besides his reservations about the RNPC, he was dissatisfied with rules enforcement in the park "particularly in the matter of liquors, etc. and the hours for automobile travel." Mather turned him down flat. The park director assured Curtis that he did not want him involved in the "wear and tear" of the superintendency but there were other reasons. Curtis was unused to the restraints of a bureaucracy and the deference to decisions from higher-ups expected of a team player. As his statement to Mather revealed, he was a stickler for park rules without much sense of the forbearance or nuance required in the difficult or ambiguous situations that a superintendent was likely to face. Neither Mather nor Albright trusted Curtis's judgment concerning a young ranger whom Curtis lauded while urging a permanent position for him at Rainier. Curtis praised the ranger's "extremely hazardous" recovery of the body of the young woman who fell into the crevasse, his "skill" in "first aid work," and his "diligence in enforcing the regulations."[15]

Albright told Curtis that the ranger was not retained at Rainier because "he had not acted properly" there. Albright agreed to give him a trial at Yellowstone after cautioning the young man that he "would tolerate no foolishness or insubordination" but the ranger resigned "after a few days' service." "I do not like the man for just about a dozen reasons," Albright asserted. Mather supported Albright's judgment of the young man in a separate letter to Curtis. "I give him all credit for what he did… like securing the girl's body…but this does not offset his other lines of action." Neither Albright nor Mather was specific about the ranger's misdeeds but their message was clear: Curtis had thoroughly misjudged the ranger. "He is certainly not a man we would want in the National Park Service," Albright concluded. At the same time Albright and Mather assured Curtis that neither had criticized Curtis, or that either believed him to be "heading a faction which was seeking to disrupt the National Park Service," as Curtis phrased it. They both valued Curtis's devotion to the park and laid rumors to the contrary at the door of the discredited park ranger.[16]

Curtis's relationship with the RNPC improved after he ended his guide work. "I did not want to go back to the company and could not afford to do so." At the same time, he wrote, the company board asked him to help sell stock and he agreed, "without hope or desire for financial gain."[17] Although he insisted to Mather that he would not mute his concerns about the company, he could not have trumpeted his criti-

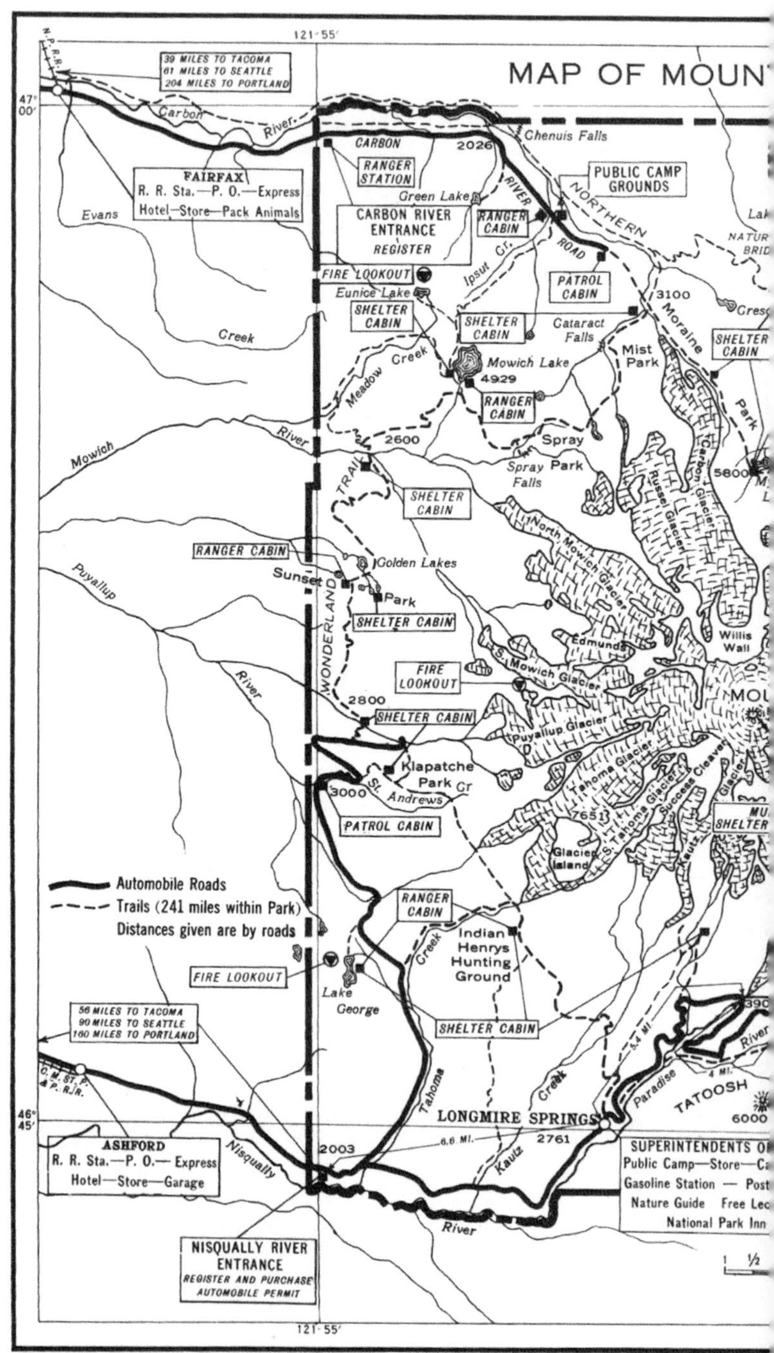

On this National Park Service map dated 1934, dotted lines below the legend PARADISE VALLEY, titled OHANAPECOSH RANGER STATION, lower right, going north to the

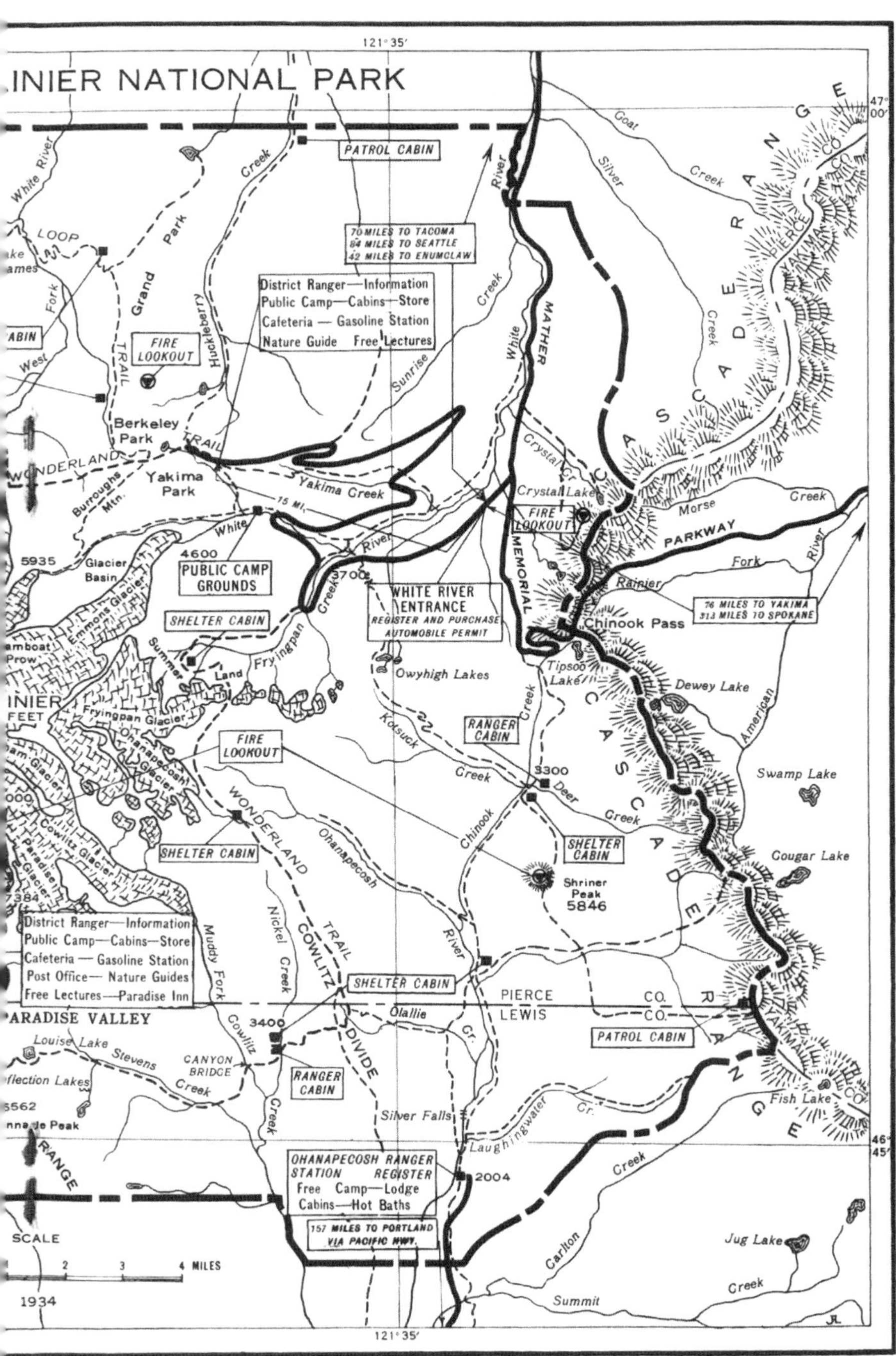

Mather Memorial Parkway lower center, mark part of the projected Stevens Canyon Road. Dotted lines from the box indicate the general location of the East Side Road. *National Park Service*

cisms while trying to sell its stock. In fact he worked with the company to improve park roads, necessary to the company's carrier business and essential to his vision of a heavily patronized park. His personal relationships with two RNPC general managers, Martin and later Paul Sceva, eased his efforts. Martin was vital to the creation of the company. Sceva was hired to be Martin's assistant, later replacing him. Martin's earlier hard work in setting up the Seattle-Tacoma Rainier National Park Committee, his diligence as its secretary, and the nearly complete congruence of their viewpoints made him Curtis's valued compeer.[18]

Curtis and Martin collaborated closely on their attempt to complete the park's West Side Road, an attempt that failed. Its failure spilled more ink and consumed more typewriter ribbon than any other issue in Curtis's career, driving him from optimism to anger, then to despair, and back to optimism. The West Side Road was a vital part of the Seattle-Tacoma committee's 1911 call for a circumferential road in the park, a visionary request at a time when the one auto road was unpaved, one-way, did not reach Paradise, and its car speed was reduced to a uniform six miles per hour. The quest for the West Side Road divides in two periods, the first from 1911 to 1924, when the divisive struggle over the proper name for Mount Rainier disbanded the RNPAB, and from 1925 to 1931 when hopes for its construction practically evaporated.

The road's construction involved flesh-and-blood people with their own visions and agendas, conceptions of the future not always dovetailing with those of Curtis and the RNPAB. They grouped into four agencies. Washington State officials stayed in the background for most of the time, and for their own reasons. Pierce County, adjacent to the park's west and north, was responsible for the road running south from the towns of Wilkeson and Carbonado, southeast to the Fairfax settlement, and on to the west boundary of the national forest, some twelve miles. The Forest Service of the Department of Agriculture had to build the road from its western boundary as it dipped southeast, then turned northeast, a total of three and a half miles to the western boundary of Rainier. The Interior Department directly, and later through the Park Service, was responsible for the Carbon River Road at the park's northwest corner. The Park Service planned a branch of the road to climb through Ipsut Pass, then turn south, past Mowich Lake, to become the north end of the West Side Road. The county, the Forest Service, and the Park Service all lacked the authority, even if they possessed the desire, to build a foot of road in the territory of any other jurisdiction. The state could build through Pierce County but not in the national forest or national park. It was, as Martin wrote in 1913, "a question so marvelously complicated" that it seemed to defy untangling.[19]

Pierce County early agreed to spend money on its portion of the road. The first stumbling block was the attitude of Chief Forester Henry S. Graves, who believed in the value of the completed road to Rainier, but believed more strongly in a road designed to help settlers and lumber producers, not city folks and tourists. Samuel Lancaster, the inter-city committee's lobbyist, made a little headway when he interviewed President William Howard Taft concerning the issue. He succeeded in persuading Taft, a Rainier enthusiast, to write Graves about improving the forest road. The rotund president recalled the "perfect torture" he endured on the forest road during his 1911 visit to the park, and asked Graves to macadamize it. Lancaster's persuasive power with Taft, Graves, and others, jarred some money lose from Graves, if not sufficient funds for paving, at least enough to make the road passable. By April 1913 Curtis believed "that the last tangle in connection with" the hookup to Rainier "is straightened" and that "it now looks as if we would have a road ready for the tourist season this year."[20] His forecast was wildly optimistic.

Road building stalled, but Curtis, Martin, and Mather continued to offer encouragement. In April 1921 the Forest Service and Pierce County agreed to push forward on their respective roads. All the effort was for naught. In June 1923, the three Pierce County commissioners reneged on their part of the bargain. The reasons may have included a lack of funds; a conviction that the county was called upon to invest too much in one road; a determination to block the development of the northwest entrance because it bypassed Tacoma, which Curtis believed; or all of them. Bad as the Pierce County-Forest Service roads were, a *P-I* reporter drove an Oldsmobile from Seattle, past Fairfax, and into the national forest before "downed timber" stopped him a half mile from the park. Even if the roads were cleared of windfalls and deadfalls, they were far from able to bear normal traffic. Worse, the Carbon River Road inside the park, the link to the West Side Road, was unpaved and "being only an earth road will undoubtedly go to pieces" if cars could get through. Curtis convinced Tomlinson, the park superintendent, to close the road before the season of 1924.[21]

By then Seattle and Tacoma were locked in a vicious struggle over the name of the mountain in their recreational back yard: should it be Mount Tacoma or Mount Rainier? The issue simmered in subdued restlessness, like the namesake volcano, only occasionally erupting in frenzy. In January 1914 Curtis could write his erstwhile partner Romans about the "few ratty individuals in Tacoma that insist on calling it Mount Tacoma, but most of them now call it 'The Mountain', without giving it a name."[22] Less than ten years later, Curtis's "few ratty individuals" transformed themselves into a determined multitude swept up in a vociferous, well-funded campaign that engulfed Tacoma and Seattle, their counties, Pierce and King, and ultimately the state, as group after group passed resolutions for retaining the name Rainier or replacing Rainier with Tacoma. Tacomans and their allies argued that "Tacoma" was a legitimate Indian name. Seattleites and their supporters claimed tradition for the long-settled "Rainier." Both assertions were challenged without dislodging them.[23]

Curtis's reasons for retaining Rainier reflected his personal situation. The RNPAB collapsed early in 1924, a direct result of the rancor over the proposed change in name. Disbanding the advisory board destroyed his work of years, perfecting an organization designed to present a unified set of national park recommendations from interested communities and groups. "Citizens of this state have been endeavoring for sometime to quiet sectional feeling and work in harmony for the development of the State," he wrote Congressman Albert Johnson of Hoquiam, who introduced a name-change resolution in the House of Representatives. "Your actions have destroyed practically all of the structure that has been built with so much care during the last two or three years." Curtis believed that "Tacoma's whole fight for the change in name is purely an advertising stunt." He accepted the argument that Native usage did not justify the name change. Not only was "Tacoma" inappropriate for the mountain, adopting it would require a corresponding change in the name of the park itself, rendering previous park advertising pointless, as well as requiring a massive overhaul of geographies and maps.[24]

Curtis's stand on the name issue aside, his bitter letters to Clarence C. Dill for introducing the Senate resolution, and to Wesley L. Jones for missing the vote on the Senate resolution, passed the bounds of reasonableness. He threatened them and Johnson with defeat at their next elections. To Dill he wrote that "you and you alone are responsible" for "your destructive acts" for which the citizens of Washington would return "you to private life." In a later letter he told Dill that "only 17% of a majority of the voters of the State voted for you," at his last election and "about 15% are sorry they did," facts that he asked Dill to remember while "casting over your future plans." Curtis's percentages made no sense, as Dill informed him, while accusing him of being "abusive" and descending to ad hominem attacks. Dill was correct, for Curtis's letters were unnecessary and immature. He should have been content with a careful statement of his views to each man, building his own reputation for probity and leaving the door open to future cooperation with all three. His wrathful letters were an unwise departure from the conciliatory stance he could assume in other situations, as during the first Olympic national park struggle and the repeated attempts to build the West Side Road.[25]

The name controversy began with a bang but ended with a whimper. In April the Senate passed a resolution favoring the change to "Mount Tacoma." The measure went to the House Committee on Public Lands where, in January 1925, the committee voted against sending it to the full House, effectively killing it. The work of Congressman John F. Miller of Seattle, the intervention of Washington, DC, lobbyist J. J. Underwood, and a letter writing campaign by Curtis and others probably had some effect.[26] In the aftermath, inflamed tempers calmed on both sides of the imbroglio. The RNPAB was reborn. Congressman Johnson and Senators Dill and Jones were reelected.

Meanwhile Curtis continued to struggle for a completed West Side Road, presenting himself in his letters as a private citizen or, later, as the chairman of the Rainier National Park Committee of the Seattle Chamber of Commerce. Before the name controversy clouded his horizon and after the clouds lifted, he tried to appeal to antagonists in a way that would release the springs of action and move the work forward. His approach demonstrated how careful and circumspect he could be, despite sometimes caustic or despairing comments to his allies in the fight. Hazel Wood Niendorff worked with Curtis at the Seattle Chamber of Commerce from the 1920s. "He understood organization procedures," Niendorff wrote, "and had a rare understanding of people and their interests which made it possible for him" to achieve the coordination of "departments of government as well as community groups and organizations." His goal was to penetrate "the interlocking organizational structures" which "he whimsically called the 'wheels within wheels' that worked toward the common objective,"[27] or should have. His understanding and his patience would be sorely tried in the ensuing years.

Curtis's unsolved problems with linking the necessary roads together worsened when Henry Ball, the new chairman of the Pierce County commissioners, early in 1924 announced that, for the second year, no funds would be spent on the county's section of the road. Moreover, he declared, no money would be spent for the next two years. It's probable, though not provable, that Ball's announcement was related to the name controversy, because the county road would be heavily used by Seattle residents, would bypass Tacoma, and would have little value for Pierce County south of Carbonado because few people other than vacationers would use it. Curtis, mindful of the name controversy, counseled a calm response from Seattle, but he reckoned without the Seattle Sportsmen's Association.

On Sunday, April 6, 1924, a faction of the sportsmen responded to Pierce County's defunding by organizing a work party of more than a thousand on the road. It followed that effort with another Sunday work expedition. Commissioner Ball was not pleased. When Curtis tried to defuse the situation in Seattle, the activist sportsmen informed him that they had sent a petition to Congress to "force Pierce County to open" the road. Curtis wondered "why they did not send a petition to God also as he has more to do with it than Congress." He enlisted the help of the respected Howard Hanson, then the deputy prosecuting attorney for King County, to convince the activist sportsmen that only the Pierce County commissioners had the legal authority to build the road. At his request, the president of the Seattle Chamber of Commerce and the secretary of the Automobile Club of Western Washington wrote to Ball, denying any responsibility for the sportsmen's outbreak. While Curtis calmed the radicals in the sportsmen's association, Dr. Horace Whitacre of Tacoma, a member of the RNPC board, persuaded the mollified Ball to find enough money to improve Pierce County's road. Early in May Curtis spoke with the circumspect president and the maverick leader of the malcontents in the sportsmen's association. Ball's action "seemed satisfactory to them and I anticipate that our troubles are over."[28]

Then the discontented faction in the sportsmen's group upset Curtis's carefully-loaded apple cart. About two hundred men returned to do more work on the road despite Ball's promise to fix it, and despite his threat to arrest anyone who attempted any more private improvement. Somehow the men escaped his notice, but Ball exploded when he discovered the sportsmen's sally. He declared that he had stopped county work on the road, and that if further private improvements were made, the road would be "absolutely boarded up and blocked to traffic of any kind." Whitacre condemned Ball for his "vacillating pettiness," and, equally, the troublemakers in the sportsmen's association. But Seattle interventionists would not be downed. The Women's Commercial Club of Seattle became the next organization to stir the pot when it sponsored a road repair trip and afterward announced the road's completion. A day later Curtis traveled the road with Mather, Senator Jones, Congressman Miller, and others, found the road in poor condition, and publicly contradicted the assertions of the Women's Commercial Club. The club president telephoned Curtis to dispute his claim. Curtis "explained to her that, in view of the fact that our cars were down to the hub in about twenty places, that I could not regard it as a highway." Mather diverted Ball from another splenetic reaction against Seattle meddling by turning his charm on the commissioner at a dinner in Tacoma, "appreciating" Pierce County's problems "and suggesting ways of working the problems out on the" county's road.[29]

Curtis took little comfort from Mather's saving the situation. During the past year, he wrote, "I have occupied the position of the rope buffer that hangs on the side of the tug and gets pretty severely bumped." Turning to a phrenological metaphor, he related how he "tried to take humorous views of the situation but on several occasions my bump of humor has been worn away." His only ill-considered public outburst came during a Seattle Sportsmen's Association meeting when he was "accused of delaying construction on the Carbon River road because I was a stockholder in the Rainier National Park Company." Although in reply he "probably said more than I should," he did receive an apology from the president of the sportsmen's association.[30]

Curtis foresaw an end to the severe bumping when Congress denied Tacomans' hopes of renaming Mount Rainier. The national defeat induced renewed cooperation between Seattle and Tacoma, but it was only one straw in the wind. Mather soothed Ball, the Park Service decided to suspend work on the Carbon River Road until Pierce County acted to finish its leg of the connecting highway, the Tacoma Chamber of Commerce reasoned with the Pierce County commissioners, and interested people in the Rainier area began looking at the possibilities of a completed road on the east side of the park, far away from Tacoma. Taken together, these developments were decisive. By early March the commissioners agreed to finish their portion of the road "within 90 days and this will enable the [federal] Government to go ahead with the Carbon River [road] work." During the same month the Mountaineers, the Seattle Sportsmen's Association, and the Women's Commercial Club agreed on a road development program for the park, an agreement settling any disputes among them or with the government agencies involved. When a Mather assistant, Arno B. Cammerer, visited in May, a conference between him and some members of the old advisory board agreed to revive the RNPAB. The next month a meeting in Tacoma ratified the decision. There was some discussion of including service clubs such as the Kiwanis, Rotary, and Lions in the new organization, but community-based groups were represented instead by their chambers of commerce or commercial clubs. Once again Curtis's "steady and guiding hand" was "at the helm" as chairman of the board.[31]

Then a bomb exploded in the midst of this happy organizational harmony. Curtis helped throw it. On August 15, 16, and 17, Curtis, two engineers from the BPR, and Thomas C. Vint, the assistant landscape architect of the Park Service, examined the West Side Road from the end of the Carbon River Road on the north to its junction with the Nisqually Road on the south. They concluded that the original survey connecting the Carbon River Road with the north end of the West Side Road was practically impossible. A road based on the survey would branch from the Carbon River road below an elevation of 2,000 feet, rise up a creek bed to 5,000 feet at Ipsut Pass, where a 1,100-foot tunnel would carry it to a 700-foot drop over a short distance to Mowich Lake. The topography required multiple switchbacks, enormously expensive and difficult to build, not to mention the tunnel. Besides, the road would slash through land almost untouched, scouring it and seriously compromising scenic views in that part of the park. All four men agreed on the unacceptable expense and aesthetic damage, recommending that the work on the West

Side Road be shifted to the south end, where no such problems existed. If the perilous branch from the Carbon River Road were not problem enough, the Carbon River Road itself was in jeopardy from river flooding. The Park Service decided to seek another entrance from the park's west boundary to Mowich Lake. The decision undermined all of Curtis's determined effort to link the Carbon River Road to outside highways, but he agreed with the change. The Carbon River Road made sense only as a route through Ipsut Pass, and with Ipsut Pass abandoned, the road had no role beyond that of an elaborate stub.[32]

The repercussions were immediate. Forsaking the Carbon River Road as a connector to the West Side Road meant that Pierce County's approach road, south to Fairfax and beyond, built at a cost of at least $600,000 after so much contention and negotiation, was rendered almost worthless to the county. The Forest Service road from the northeast corner of the national forest to the park's boundary, costing at least $60,000, suffered the same fate. These costs were substantial sums at the time and were incurred less to help Pierce County and the Forest Service than to assist the Park Service, which had repudiated its own road survey and construction, thereby radically devaluing the cooperative efforts of Pierce County and the Forest Service. The county and the Forest Service were understandably reluctant to build any part of a new approach road which had to leave the existing road at a point north of Fairfax and enter the park to the south of the Carbon River.[33]

Curtis and others at once began a campaign to have the new approach road and some connecting mileage to the north of it declared a state highway. Curtis wanted to relieve Pierce County of the burden of building its segment of the new road, but the legislature refused because the existing, underfunded state system was incomplete and practically all pressure to extend it had to be resisted. Meanwhile Curtis and others successfully appealed to the state highway commission to bring the new approach road into the forest highway system under a cooperative program with the federal government. The Forest Service agreed to begin clearing its part of the new road but funding limitations in the Pierce County segment encouraged other solutions. In February 1929 Superintendent Tomlinson asked Albright, director of the Park Service after Mather, if federal funds could be used to build the road. Tomlinson's argument was that the service had created the problem and therefore it should help the state build the new approach road. Albright passed the buck to Congressman Cramton, who replied with a firm "No." National park funds were "heavily encumbered" for "strictly national park projects and it would be very dangerous for us to inaugurate any policy of constructing approach roads." Simultaneously Curtis lobbied hard for a compromise in the legislature. If the legislature would not extend the state road system, would it instead appropriate funds to help build the approach road as the forest highway approved by the highway commission? The legislature responded with an appropriation of $200,000 for the biennium. Governor Hartley promptly vetoed it.[34]

Then H. A. Rhodes, the president of the RNPC, stepped in to try to save the situation. Rhodes suggested to Everett G. Griggs, president of the St. Paul & Tacoma Lumber Company, that the company build a logging road from the Northern Pacific track at Fairfax to the boundary of the national forest. Rhodes suggested that the logging road run over the survey line of the new approach road. In April Curtis called a meeting of several interested groups to explore the possibility. Under the arrangement the lumber company would build the road, allowing construction material to be hauled over it while logging proceeded. When the company's property was logged off, it would take up its rails and deed the right-of-way to the state. Griggs was amenable, provided that some financial assistance came from the state, the county, and the Forest Service. The plan foundered on a 6 percent grade over part of the road, in Griggs' words, "impossible for safe logging operations." A 6 percent grade on stretches of the highway was no barrier to car travel, but any grade steeper than 5 percent was a challenge to a railroad. Martin speculated that Griggs had another motive, to avoid "criticism from the public" about cutting timber in full view of the right-of-way when he could later remove the trees by another, less obvious route. According to Martin, Griggs wished to avoid the complaints which could arise when later park-bound vacationers viewed a denuded landscape. Whatever Griggs' motive, his withdrawal killed all hope for beginning a new approach road in 1929.[35] By then both Cramton and the Park Service were fed up with the state of Washington's inaction. In April 1928 Cramton had written Curtis that he was "extremely disappointed" in the state's refusal to build its part of a new road. Cramton's message was clear: there would be no more than nominal road construction on the west side of the park until the state agreed to do its part. Several months after Hartley's veto there was more bad news from lobbyist J. J. Underwood. Concerning the West Side Road, Underwood wrote, "Horace Albright told me the other day that he was beginning to think it was a mistake ever to start it. It will probably be ten years before it is finished."[36]

The ensuing Great Depression all but doomed the West Side Road. Its demise had a deep background. In 1931 Governor Hartley signed a bill appropriating funds for the new approach road through Pierce County. His about-face from his line-item veto of the same measure in 1929 was symptomatic of his Depression-induced retreat from a commitment to state government frugality. When the public's delight with his rhetoric withered in the face of mounting unemployment and despair, and his political prospects dimmed, he struggled to appease former adversaries. His road gesture came too late because the same Depression that undermined his future seriously weakened the RNPC. The company, never especially robust except during the 1920s, could not muster the capital to build a hotel at Spray Park. Spray, a long-time hiking destination located in a relatively level meadow and woods located south of Mowich Lake, is a sublime area of Rainier and the logical place for a hotel after the West Side Road was finished. Even if the RNPC could build a hotel, though, the West Side Road had to wait on road building to Rainier's boundary. The West Side Road and its spur to Spray Park could not be completed, or even seriously contemplated, until a new approach road through Pierce County and the national forest permitted a rapid construction attack on the West Side Road from the north end. That was Park Service dogma.[37]

Other road and recreational developments further dimmed the prospects for the West Side Road. In 1930 the Park Service finished a road to Sunrise, then better known as Yakima Park, in Rainier's magnificent northeast corner. The company was no more able to build a resort hotel at Sunrise than it was at Spray, but in 1931 it completed a lodge and cabins that together would house about as many vacationers as a hotel, in addition to the usual staff and services. Until then the partisans of the West Side Road endlessly repeated their litany—the completed road was essential because it would siphon visitors away from overcrowded Paradise. The Sunrise development helped to take the pressure off Paradise, but by July 1933 Paradise needed no help in depressurizing. July weekends, the highest demand dates of the season, saw vacant rooms and an empty dining room at the company's Paradise Inn. Nor did an attendance boom at Rainier after 1933 rescue the beleaguered RNPC. Depression-era economizing encouraged visitors to stay in auto camps, not the company's hotel at Paradise. An uncrowded Paradise and the RNPC's inability to develop Spray Park were powerful arguments against the West Side Road.[38]

Whatever the fate of the West Side Road, Park Service officials were in any case backing away from elaborate road-building schemes in Rainier. The huge expense of road building and a rising interest in "roadless" and "wilderness" areas prompted their change of course. The first to go was the unconstructed "loop" road around the mountain, a conception dating from the park's early days. Unfortunately for the loop road, the north boundary of Rainier was drawn too close to the mountain's flanks for a practical north side road to surmount their heights, then plunge into their declivities, only to rise again. When Curtis and his fellow analysts studied the situation on the ground and rejected the road through Ipsut Pass, they severed the only possible link between the north side road and the original Carbon River Road-West Side Road connection. Curtis, who knew the mountain's north face well, in 1925 doubted "that a road could be constructed across the north side." The next year the Park Service eliminated the north side road from its plans. "It may be," Curtis wrote in 1927, "that sometime in the dim distant future" a north side road would be pushed through, but his phrasing encouraged skepticism. In 1928 Mather decided the issue, declaring the North Slope a wilderness area, sequestering it from roads or from anything but the barest future development.[39]

Ultimately, Congressman Cramton and Park Service Director Albright were willing to delay the West Side Road for a long time. Both men were acute, and surely understood the attitudes of the governor and the legislature, as well as the reluctance of Pierce County and the Forest Service to rush into spending fresh funds on new approaches when they had sunk large sums into roads that the Park Service had rendered practically useless. It would be simple to condemn Cramton, Albright, and the Park Service for knowing the state's problems yet blaming the state for delaying the road, but the criticism would be misdirected. The improvement and maintenance of the national parks, and especially of their roads, involved great sums which, in the nature of things, were never enough. It was easier to postpone significant work on the West Side Road than to put money into a road and hotel venture that might never be fully realized.[40]

To put the problem another way, park officials at first thought of a circular O road around the mountain, then of a U-shaped road system without the north side road, and finally a system in the shape of a reverse capital L, running west to east on the park's south side, then vertically on the east side south to north, with a southerly extension to tap the southeastern corner. They never articulated these dynamics, partly because a wait-and-see attitude toward the West Side Road did not force any decisions. When the new approach and intra-park roads to Mowich Lake opened in 1935, the company's inability to build a hotel at Spray was all too evident.[41] Without a developed tourist destination there was no need to extend the West Side Road south of the lake. In 1936 Curtis asked Cammerer, by then the Park Service director, for funds to pave the road. Cammerer brushed him off with four paragraphs of bureaucratese.[42] Curtis's "wheels within wheels" never turned a full revolution.

Curtis failed to further the West Side Road, one of his major Rainier preoccupations. But he cannot be blamed for that, because he had no control over the weak RNPC, a poor road survey up the Carbon River to the impossible Ipsut Pass, or the economic calamity of the Great Depression. He did what he could do: seek the cooperation of all interested parties. His great triumph was the RNPAB, organized in cooperation with Thomas H. Martin. The advisory board played a significant role in Rainier's development, in no small measure because its membership included the leading commercial organizations of the communities surrounding the park. The board's commitment to unified recommendations to the Park Service muted rivalries within its constituent groups and made its suggestions all the more compelling.

Chapter Eight

Mount Rainier: Developing the Park, Dissolving the Rainier National Park Advisory Board

Curtis's primary goal through the RNPAB's later existence remained, as ever, advancing all Rainier interests in harness, while working toward improving the park. Lobbyists played a pivotal role in retaining the interest of Washington, DC, officials, and keeping them posted on local concerns. They also informed Curtis and the RNPAB about thinking in the national capital. Curtis's continuing association with the Park Service, nationally and locally, was usually smooth though it involved him in a contentious road location during 1931. He assumed a leading role in establishing the Mather Memorial Parkway in 1932. The event honored Mather, of course, but it also drew the Park Service and the Forest Service together to plan the commemorative road. The RNPAB cooperative structure, so carefully built, collapsed in 1936 when Curtis published an article opposing the creation of the Olympic National Park. That debacle was, fortunately, in the future and unimagined while Curtis advanced his cooperative agenda.

Curtis and the advisory board had good success with their ultimate choice of lobbyist, though at the beginning their later fortune was not so apparent. First they hired Samuel C. Lancaster, an inventor, engineer, preservationist, and author whose first love was road building, not lobbying. Nevertheless he went to work with a will, intervening with President Taft on behalf of the original Forest Service approach road, and giving several successful slide-show talks on Rainier. His sincerity and dedication failed to prevent a deep cut in the park's Congressional appropriation once it reached the Interior Department. Overall, Lancaster's four months' work from January to April 1913 made no more than a modest impact.[1] The next lobbyist was General James A. Drain, later a national commander of the American Legion. In the summer of 1914 Drain's retainer was exhausted, again with little result.[2] Informal contacts filled the lobbying void until the early 1920s.

The outlook brightened considerably when J. J. "Jack" Underwood, the manager of the Washington Bureau of the Seattle Chamber of Commerce, became, in effect, the lobbyist for the advisory board. Underwood, a former newspaperman, was well-informed, industrious, genial, and gregarious. He was also persistent. He thrived on ingratiating himself with the national parks administrators, members of Congress, and functionaries in the executive branch. Always he pushed for greater Congressional appropriations for Rainier and the related though separate issue of increased spending authorizations from the executive branch. On at least one occasion he wrote the Park Service's report on its appropriation bill, while at the same time he persuaded Albright to increase the allotment for Rainier. On another occasion he gained an appointment with President Calvin Coolidge, successfully enlisting presidential intervention to increase a Congressional appropriation. His letters and copies of letters about his activities flowed to Curtis, Martin, the RNPC, and, of course, to the Seattle Chamber of Commerce.[3]

Curtis praised Underwood's essential contribution. "We must all remember that it was through your efforts that we put over the National Parks bill as far as money is concerned," he wrote Underwood in 1930. Underwood did not work alone, Curtis conceded, "but we would not be anywhere if it was not for the money and you got the money."[4]

Curtis never relaxed his own labors. Much of his effort went into keeping the Park Service, the RNPC, and the advisory board working in harmony. Except for occasional bumps, ultimately smoothed over, his struggles succeeded in locating the "wheels within wheels" and keeping them turning. One dispute concerned funding renewal, when Albright, not noted for his measured responses to criticism, accused Curtis of having "inadvertently done me a great injustice" by complaining to Cramton and others about delayed work on Mount Rainier roads. Curtis answered Albright calmly, pointing to his devotion to the national park idea, the advisory board's ability to go outside channels to increase appropriations, the board's lack of "desire to seek any credit for the work we are doing," and his wish neither to criticize Albright nor deprive him of credit for his "splendid" service. "I might call [Albright's] attention to some very serious inconsistencies" in his letter, Curtis wrote Tomlinson, but "I have no desire to enter on a long correspondence controversy." The dispute blew over. Occasionally Curtis moved matters along by accepting the advice of others to moderate his efforts, as he did in 1928, when he instead wished to press for increased funding for all the national parks.[5]

Park officials and the BPR responded to Curtis's obvious interest in advancing Rainier by asking for his advice and the advice of the RNPAB on numerous occasions. They well knew that his interest in developing the park did not suggest disinterest in how it was done. Park roads were the usual subject of those requests, although trails were sometimes included. Excepting one issue, the Park Service usually agreed with Curtis's or the advisory board's conclusions.[6]

That one issue surfaced in 1931. It involved the extension of the South Side Road from east of Paradise to a connection with the East Side Road. Curtis was reluctant to involve the advisory board or himself in specific road locations because he recognized that choosing a route entangled the chooser in engineering details. But there were precedents. Secretary of the Interior Franklin K. Lane had asked the old Inter-City Committee for help with route selection. The superintendent of Rainier had asked Curtis to make an early reconnaissance of the park's west side. The Park Service had asked him to participate in the Carbon River-West Side Road reconnaissance and to help write the report giving up its Ipsut Pass route. Besides, a lot was at stake. Curtis defined the purpose and development of national parks "to be for the enjoyment and benefit of the people," and for "recreation" through "visits to our shrines of natural beauty" which "will make a better citizenship." His sentiments were conventional but his eagerness to move people in cars through the scenic high country was controversial. The proper route, as Albright summarized it, centered on "a conflict between the points of view of the hiker and horseback rider who want all the high country completely reserved and the motorist who thinks he ought to be able to get everywhere…in his automobile."[7]

Albright exaggerated the motorist's desire because he was writing to an advocate of reserving the high country for hikers and horseback riders. Certainly neither Curtis nor the members of the RNPAB wanted unrestrained auto access to all parts of the park. They did want opportunities for the drivers and passengers in a car to see Mount Rainier and other impressive features. By their calculations 95 percent of park visitors arrived by car and never walked far from the road. A small but active group of preservationists opposed giving in to what a later park ranger would term the "indolent masses," the car-bound majority and its sedentary predilections. Preservationists wanted the upper areas and their views limited to those who were willing to forsake comfort and hit the trail. At first Curtis favored a route, perhaps including the Stevens Ridge, up Nickel Creek to the Cowlitz Divide, a high ridge along the divide north to the vicinity of Indian Bar, pivoting southeast along the Ohanapecosh River (until the river turned south), then to Chinook Creek, then north up Chinook Creek to the vicinity of Cayuse Pass. The preservationists wanted a road southeast down the Stevens Creek-Cowlitz River Valleys, then looping north up the Ohanapecosh-Chinook Rivers to Cayuse Pass. Curtis denounced the preservationists' route because it ran almost entirely through heavily timbered country and offered practically no views.[8]

A blizzard of mail fell on Albright's desk, correspondence favoring either the high country or low country road approaches. In May Albright split the difference for part of the distance and deferred a decision on the rest. He ordered a road benched in along Stevens Canyon, above the creek but below the ridge, where it is today. He asked the BPR to survey the possible routes from the Stevens Canyon Road once it emerged from the canyon. The survey quashed Curtis's hopes for his projected road. Tomlinson, who also favored Curtis's route or something nearly identical, wrote the sad news. Curtis's course from the north end of the Cowlitz Divide through the confluence of the Ohanapecosh and the Chinook involved "excessive grades, very poor alignment and heavy maintenance costs because of numerous snow pockets." An alternate plan similar to Curtis's was even worse from the standpoint of extra construction and overall cost. Albright continued trying to please both the automobilists and the preservationists. He planned a short stretch of road south along the Cowlitz Divide, which then turned northeast through the heavily timbered Ohanapecosh-Chinook Valley. He ran a stub road north along the Cowlitz Divide to the "high observation point near the north end of the Divide," where it dead-ended. After Albright apportioned the road-and-trail baby, Curtis and the advisory board beat back an attempt to move the South Side Road outside of the park to preserve the park's southern reaches from development.

Questions about the exact location of Albright's future road remained, but the battle appeared to be over.[9]

Alas for Curtis's hopes, World War II interrupted road construction at Rainier. When the South Side Road (now the Stevens Canyon Road) was completed in the late 1950s it followed a deep valley route from Stevens Canyon to the east side, while the scenic Cowlitz Divide remained the preserve of hikers. Curtis would have disapproved. In 1933 he wrote that Albright's split-the-difference road program would leave the hiker "ample ground for his exploits. Having granted him this I feel he has gone beyond his rights in demanding that the rest of the Park be held sacred to such a small percentage of our citizens."[10]

The hiker versus auto struggle involved the Mountaineers because some members of the climbing club, though not the organization officially, fought to keep cars off the ridges. They were especially concerned about the Cowlitz Divide, where Curtis's road plans would have displaced parts of the Wonderland Trail, the circumferential hiking trail around the mountain. Though justifiably estranged from the Mountaineers since 1912, Curtis worked hard to keep the organization allied with his larger purpose of developing Rainier. Before and after leaving the club he supported their efforts to build a stone shelter at Camp Muir on the mountain's south face. In 1925 he arranged an agreement among the Mountaineers and the other members of the revived advisory board to support a plan for road development at Rainier, including the then-viable West Side Road. In 1927 the Mountaineers again endorsed the cooperative nature of the advisory board but insisted that "at all times we must reserve the right to act independently should we see fit, rather than to be bound by any majority rule." Curtis accepted the club's continuing membership because he already had their agreement to the RNPAB's road program.[11]

The next year Curtis asked Meany to appoint a committee of the Mountaineers to work with the advisory board on trail development in Rainier because of the club's familiarity with the park. Meany responded warmly to Curtis's suggestion.[12] Though Curtis sometimes expressed exasperation with members of the Mountaineers who acted independently of the organization, he strove to keep the group informed of advisory board activities throughout the life of the RNPAB.[13]

Meanwhile Curtis forged ahead with other activities related to Rainier. Nearest his heart was a nationwide movement to commemorate Mather, an effort that began following Mather's stroke in 1928 and his resignation as the Park Service director, officially on January 12, 1929. The undertaking gained greater urgency and poignancy after Mather's death on January 22, 1930. The origins of the commemoration began much earlier, in 1915. While visiting the park that year, Mather became enchanted with the idea of developing a "Cascade Parkway." In his vision the road ran from a point north of the present northeast boundary of the park, south along the east bank of the White River, then turned northeast at Cayuse Pass to follow the north banks of the American, Bumping, and Naches Rivers, finally turning southeast along the Naches. The projected road, then running entirely through Forest Service territory, now forms part of Washington State Route 410. At the time Mather was the Assistant to the Secretary of the Interior in charge of the national parks. When he became the first director of the newly formed Park Service some two years later the situation did not auger well for his Cascade Parkway, partly because "many Forest Service men" whose cooperation he would need "thought him a rich, supercilious autocrat who looked down on" them.[14]

Then events intervened. At the beginning of the mid-1920s a semi-official commission aired interservice disputes, settling them and smoothing over the antagonism between the Park Service and the Forest Service. In November 1928 Mather suffered his first stroke and was out of active management from then on. In the later 1920s the Forest Service significantly expanded its preservation and recreational activities, including land along the line of the proposed Cascade Parkway. In 1929 the Forest Service and the RNPAB collaborated on a plan setting aside a strip one to three miles wide along the road. No commercial timber cutting would be permitted in the area; trees could be cut only for firewood or cabin construction. A few lodges would be permitted but these, together with the private cabins and campgrounds, would not intrude on the roadway. Commercial timber harvesting would not be allowed in sight of the road. Signs would be limited to those "absolutely essential for the guidance of the traveling public" and the entire area maintained "in as nearly a natural condition as is possible." In 1931 Congress expanded Rainier's eastern boundary, moving about 10 percent of the road into the park and under Park Service regulations.[15]

While the actions of the Forest Service and others secured the proposed parkway from exploitation, the work on the Stephen T. Mather Appreciation, as it was called, went forward. John Hays Hammond, a talented mining engineer, fiercely determined, wealthy, and a wheelhorse in the Republican Party, chaired the group. About 275 people, including Curtis, belonged. Curtis and his allies submitted their memorial idea, converting the Cascade Parkway into the Mather Memorial Parkway. The appreciation committee received more than forty memorial proposals, of which "our project was listed down at the bottom of a long list of projects," Curtis wrote, though he later generously credited Hammond with a large role in realizing the memorial parkway. The committee, overwhelmed with proposals, at last decided on bronze plaques depicting Mather in bas relief, facing trees against a mountain backdrop, with a quotation from Cramton beneath. The plaques would be placed in all the national parks and monuments. The committee did not, however, rule out other memorial activities; indeed, it encouraged them.[16]

The Rainier celebration would include the parkway project, there was no doubt about that. Renaming the sixty miles of the newly-completed Natchez Highway fell to the secretaries of Agriculture and Interior, who cooperated in making the change. Curtis moved up the dedication of the Mather Memorial Parkway to July 2, 1932, a date that coincided with the conclusion of the "good roads" automobile International Progress Tour to British Columbia. It also closely matched other dedicatory ceremonies to be held around the country during the same month. Curtis invited his old nemesis, Governor Hartley, in recognition of Hartley's late conversion to funding the approach road to Rainier. Besides, he was the head of the state government. When dedication day arrived, the snow was deep at the head of Chinook Pass, too deep to install the bronze plaque on a boulder. Tomlinson improvised an easel to hold the plaque while the ceremony unfolded. Curtis, the master of ceremonies, introduced Hartley, who gave an address favoring an exchange of private timber for state-owned woodland, to preserve some trees near the parkway. It was a remarkable gesture for the aggressive lumberman. Meany, a fixture at such occasions, also spoke. Then Curtis called upon Tomlinson for a few brief remarks and finally, on the Forest Service representative for similar comments. Throughout Curtis kept the focus on Mather and his achievements.[17]

Curtis participated in another "appreciation" of Mather when the *Washington Motorist* devoted its April 1930 issue to Mather's life and accomplishments. In his article Curtis recalled how Mather came into the "national park system" when it "was a very obscure part of" the Interior Department. Mather changed all that with his inspiring leadership. Curtis remembered the 1905 climb on Mount Rainier, even though "there was no hint at this time that our small party held the man who would do more than any other to establish the national parks to their present high standard." The 1905 event, when Mather "slept out under the trees" and "explored the glaciers" was part of his inspiration to develop the parks. "Steve is dead," Curtis concluded, but "he will live on through generations."[18]

Curtis's veneration of Mather was sincere but it should not obscure his exasperation with some aspects of the Mather and Mather-Albright administrations. He complained about the Park Service's dallying over contracts for clearing rights of way, and for grading and paving roads. These delays cut into the short road building season, frustrating his hopes for motorist access.[19] Given his opposition to introducing non-native plants along roadsides, he could not have approved of Mather's and Albright's introduction of exotic fish, trees, shrubs, and other plants in the national parks. No record of Curtis's overt opposition survives, perhaps because most of those introductions occurred in Yellowstone or in other parks besides Rainier. Probably Curtis thought it best not to contest the policy, opposition that would create antagonism and hamper his work for Rainier's development.[20]

In sum, beyond the minor or transient disagreements among people or groups with the same goal but differing notions of how best to achieve it, Curtis got along well with Superintendent Tomlinson and his predecessors. He cooperated well with Martin, Sceva, and others of the RNPC, with key state legislators, and with Cramton. Other members of Congress, except for Dill, Jones, and Johnson during the Rainier name controversy, received polite letters from him. He worked closely with Mather, Albright, and, briefly, with Albright's successor, Cammerer. He reconciled with Governor Hartley after that worthy's conversion to highways and conservation. He even got along with the Mountaineers as a group, despite their occasional mutual exasperation.

Much more than the advisory board occupied Curtis's activities related to Rainier. He held a commemorative ceremony for his friend Simon Fraser Tolmie, then in the waning months of his premiership of British Columbia, at the uncompleted Mowich entrance to the park. The September 1933 gathering celebrated the centennial of the visit of William Fraser Tolmie, Tolmie's father, who was the first white man to arrive in the area. The first Euro-American appearance in the park's environs would not be celebrated or even acknowledged two generations later, but at the time it was considered worthy of speeches and a plaque.[21] Curtis organized letter writing campaigns on behalf of appropriations for Rainier,[22] answered park visitors who complained about one or another aspect of their experience,[23] took many photographs of the mountain,[24] wrote numerous articles and photo essays about the park,[25] privately published a Rainier book illustrated with his own photographs,[26] and gave innumerable slide show talks concerning the beauties and needs of the park and its mountain.[27] Working with the tact and diplomacy that characterized him when he had no personal stake in the matter, he helped to calm the dispute over naming the eastern resort area newly opened to auto traffic.[28] He gave time, thought, and energy to accompanying congressmen and other visiting firemen through the park.[29] He opposed grazing in Rainier.[30] In one of his occasional ventures into national park policy, he advised against diverting water in Yellowstone for irrigation. He took the preservationist stand despite his extraordinary commitment to irrigation in eastern Washington.[31]

Curtis was involved in the state parks movement, although he thought of state parks in Washington as a supplement to Rainier and the recreational areas in the national forests. In 1925 he argued unsuccessfully for a state park bond program to finance new parks but lost interest partly because of Hartley's determined opposition to any state program. "Washington has enough State Parks at the present time with a very few possible exceptions," Curtis wrote in 1929.[32] He was sporadically involved with the National Parks Highway Association[33] and the National Parks Association, as well as with the Washington State Chamber of Commerce[34] but was too caught up in other causes to give substantial time to any of them.

All the while Curtis continued his twin focus on Rainier itself and on the role of the RNPAB in advancing the park's development. In 1936 his personal interest in the park ended, although he never intended that outcome. His article of that year on the proposed Olympic National Park was the catalyst (see Chapter Ten). Just as a dustup in 1912 over an earlier proposed Olympic Mountains park caused his break with the Mountaineers, his 1936 article inflamed the later issue, swept away the advisory board, and terminated his close contact with developments in Rainier. The storm broke in April after his article in *American Forests and Forest Life* criticized the planned Olympic park as too large, and favored Forest Service over Park Service management of a federal property smaller than the proposed national park. It is important to note that the identifier, "Chairman, Rainier National Park Advisory Board," appeared below Curtis's name.

On April 15 Thaddeus A. Stevenson, the manager of the Tacoma Chamber of Commerce, dispatched an explosive letter to Curtis. "In this article you definitely take sides," he told his erstwhile compatriot, "in a matter that is highly controversial." The Tacoma chamber was a member of the advisory board, "and we object to having you, as the chairman, speak for us on this matter without our knowledge or consent." Curtis had "unnecessarily offended the officials of the Park Service" and had "assumed authority" to speak for the chamber, although the chamber had not taken any position on the Olympic park question. Indeed, the chamber had no objection to a park "with boundaries reasonable and satisfactory to the communities interested in its development."[35]

Curtis's response was unequivocal. "I regret as much as you do" the identifying line, he declared to Stevenson. "The article was not so signed and I was not presuming to speak for the Advisory Board." He had opposed the board's dealing with anything other than Rainier, and had opposed the group assuming any position regarding the Olympic park. The identifier appeared, he declared, because, after receiving Curtis's manuscript, the editor of *American Forests* "wrote inquiring what I did and I listed a number of my activities. Evidently he chose this one and used it in the heading of the article." He wrote similar letters to Cammerer and to all of the members of the RNPAB.[36]

Stevenson was not appeased. He suggested to Curtis "as tactfully as I could, that he resign as chairman…in the interest of the Park." For the rest of 1936 and into 1937 Curtis did not call any meetings of the advisory board. During that time he received indirect, and perhaps direct, oral comment from Tomlinson about his article, his diminished standing with the Park Service and its parent Interior Department, and the insistence of the Park Service on changing the name and membership of the board. In February 1937 he decided not to resign his chairmanship because the Park Service would not recognize a successor, effectively nullifying the existence of the RNPAB. Instead he chose to "close the affairs" of the advisory board. A last financial statement listed an "Unpaid Balance" of $63.01.[37]

The advisory board's demise leaves several issues to be sorted out. The first is Curtis's justification for his identification as chairman of the advisory board at the head of his explosively controversial article. Simply put, there was no basis for his statement that the article "was not so signed." Two surviving copies of the manuscript with title page intact include, beneath his name, "Chairman Rainier National Park Advisory Board." The title that Curtis himself supplied differs only in an added comma from the identifying line in *American Forests*.[38] There is no correspondence in the Curtis papers between Ovid Butler, the editor, and Curtis to corroborate Curtis's claim that Butler chose the advisory board chairmanship from a list of Curtis's activities, then included the title in the article. There is no indication that Curtis objected to the title, "Chairman Rainier Park Advisory Board" at any stage in the article's preparation for publication, or after its publication. Had there been any correspondence of the sort Curtis surely would have presented it to buttress his claims of innocence. Blaming Butler for the wording was shabby. Curtis was on firmer ground when he expressed his "regret" and insisted that he was "not presuming to speak for the Advisory Board."[39] The best that could be said for Curtis in this episode was that he added the "Chairman" line to buttress his authority to write on national park matters without reflecting on its implications.

The second issue involves whether Curtis should have written the article in the first place. Here the answer is, probably not. His views on the Olympic Park issue were well known and already limiting his effectiveness with the Park Service. Had he insisted on writing the article he should have resigned as chairman of the advisory board before its publication, or, indeed, before he wrote any negative comment about an Olympic park. Even in the article's bitter aftermath he could not understand how his opposition to one national park in Washington State undermined his usefulness in developing the other. He could not grasp why Park Service officials, equally committed to both Rainier and Olympic, would be offended when he attacked a park plan which they were working to realize. He was simply too emotionally and intellectually caught up with the Olympic park issue to recognize the contradictions in his position.

Finally, should his critics have been as harsh as they were? Again, the answer is no. Stevenson's fiery letter contained the phrase "you have unnecessarily offended the officials of the Park Service," an unmistakable suggestion of a prior conversation and agreement between Tomlinson or others in the Park Service to criticize Curtis. Later Stevenson sent Curtis a gracious letter while in the same missive strongly suggesting his resignation.[40] The Park Service may have been searching for a way to oust Curtis, although a more temperate approach would have been to encourage Curtis's resignation while marginalizing the advisory board under a more compliant chairman. Calm judgment is, however, often in short supply in emotional situations such as the one of 1936.

The unfortunate parting of Curtis and the Park Service should not obscure his significant role in the sustained development of Rainier. His record of settling disputes, advancing reasonable road development, turning aside ill-conceived notions such as the road through Ipsut Pass, and advertising the park through varied publicity, cannot be overemphasized. Moreover, contemporaries valued his work highly, and their unsolicited testimony spoke volumes about their affection and regard for the doughty photographer. In 1924, at the height of the name controversy, Martin responded to an attack on Curtis from another Tacoman by writing "that Mr. Curtis...has probably done more in a voluntary way for the development of the Park than any other man in the state; and he has no compensation for this service."[41] Three years later Martin thanked Curtis for a photo portfolio. "Your painstaking and intelligent work in all matters relating to the Park is truly wonderful."[42] In 1929 Tomlinson acknowledged a "fine set of pictures" which "did more than anything else to convince the Director and his assistants of the need" for more park funds.[43]

Other correspondents generalized from Curtis's Rainier activity. One wrote about how "you were rather careless about the credit but you were very much interested in the accomplishment."[44] Although they would later part company, Stevenson of the Tacoma Chamber of Commerce declared that after six years of observing him, "I do not think there is a man in the State of Washington that has labored more courageously and more efficiently in the interest of the public than you."[45] When Curtis wrote a letter regretting Albright's retirement from the Park Service, he received a generous reply. "I always enjoyed my association with you," Albright wrote, tactfully overlooking some occasional contentions. The letter acknowledged his and Mather's indebtedness to Curtis's "interest and active support," a "local support in greater measure than we have obtained anywhere else in the country." Both Mather and he "have been profoundly grateful to you."[46] It was a fitting coda to Curtis's work for Rainier.

CHAPTER NINE

The Olympics: Climbing the Mountains, Photographing the Peninsula

ASAHEL CURTIS knew the Olympics.

The tenth of August 1907 dawned cloudy and cold. Mist and a rising wind heightened the chill and the gloom. Curtis led forty-five men and women on the first annual outing of the Mountaineers as they filed out of a temporary camp deep in the Olympic Mountains. They planned to climb the namesake Mount Olympus after hiking almost seventy miles from the town of Port Angeles on the Olympic Peninsula's north coast. Their first sixty miles were relatively easy because a trail was prepared. The last eight miles were tougher, and now the going would be tougher still, with no trail, a steepening climb, and the increasing threats of altitude sickness and disorientation in the frigid murk.[1]

Mist became rain while the climbers tramped upward along the Humes Glacier of Olympus's east face, maintaining their "company formation," the military style of organization inherited from the Sierra Club and the Mazamas.[2] Curtis and his "general staff" took the lead, halting the column whenever a member became dizzy or short of breath. The cases of altitude sickness slowed the climb but the group's safety was paramount. All would ascend together. No one could be left alone to recover because the cold could claim them or the clouds envelop them beyond any hope of rescue. For the same reasons nobody could be sent back to the temporary camp because of the dangers facing a lone Mountaineer making a treacherously rocky descent.

Soon the rain changed to snow and the wind continued to rise. A thirty mile per hour blast pummeled the group when it emerged from the protection of a ridge. Waterproofed boots, rubber slickers or oilskins, and heavy woolens were not enough to deflect the gale. Worse, the clouds hid the next necessary step to the summit, a 700-foot drop to the Hoh Glacier, forbiddingly steep even on a clear day. Clouds wrapped the three crowns of Olympus. The West Peak, the highest, the farthest, and the most worthy of ascent, towered at least three miles beyond.

"The party appeared suspended in the heavens on the edge of some great cloud, with a white desolate world forming out of the chaos," Curtis wrote.[3] Three climbers "were exhausted," and the August weather was impossible, so no lives would be risked "for the mere bauble of a mountain summit."[4] The man to whom that "mere bauble" meant much commanded his captains to form into companies, they gave the necessary orders, and the bugler—yes, there was a bugler—sounded Retreat, the companies managed a "faint cheer,"[5] and tramped downward through the swirling snow.

One of the forty-five climbers suffered a bad fall and serious injuries. She was with three other women attached to a company descending to the temporary camp when her group became separated from the company. Visibility was near zero. While she and her companions groped for an alternate way down, she slipped, fell one hundred feet over rocky terrain, and landed at the bottom of a cliff, exhausted, badly cut, bruised, and in shock. "That the fall was not fatal seemed miraculous," Curtis declared.[6] A doctor with the party reached the victim "within five minutes." The snow returned to a "driving rain" and the only protection for doctor and patient was "a strip of canvas stretched against a rock." The doctor arrived prepared, however, dressing all "her wounds perfectly, even though one cut alone required eleven stitches."[7]

The injured climber recovered under the care of another doctor, Curtis's friend Eaton. Meanwhile about two-thirds of the party, including Curtis, returned to Seattle. Eleven stay-behinds reached the West Peak of Mount Olympus on August 13, in marginally better weather than three days before. Previous claims notwithstanding, on their second try the Mountaineers made the first recorded ascent of the highest peak in the Olympics. The party's barometer recorded its height at 8,250 feet, which more precise methods later reduced to 7,965. Assorted members of the original group and the conquerors of the West Peak also climbed seven other mountains, as well as the two lower peaks of Olympus.[8]

Curtis superbly handled the desperate Olympus situation, basing his pivotal decision to abandon the climb partly on his grasp of group psychology. The members of his party started out "seemingly very cheerful and happy" but after rain changed to snow and the wind rose they were reduced to "hoping against fate." At the head of the Humes Glacier they felt the "full fury of the gale and everyone's hopes sank." He could have led the discouraged party onward because, as another Mountaineer phased it, "the discipline was absolute." A decision to go forward in the face of looming danger and low morale would have risked lives and threatened the system of strict obedience. Group cohesion aside, the weather doomed Curtis's grasp for the "mere bauble" of the mountain top. Once he decided on the only "reasonable course" of retreat he could not wallow in disappointment, for he faced challenges more demanding than those of the ascent. Descending, he had to lead from the front, deciding on the details of the return route. On his authority "a small party was ordered from the line and sent in advance to start fires and prepare for the main party." It was a risky decision but taken for the comfort of the majority. He organized the rescue of the injured Mountaineer, making the final decision to leave her behind under Eaton's care. The decisions to direct the "main party" to the base camp the same day, and to begin the return march to Port Angeles the following morning were his.[9] Curtis would not have made all those decisions without advice from trusted members of the group, but the consultations did not relieve him of the ultimate responsibility.

For all of his decisiveness and bravery, Curtis made an unprepossessing leader. A photograph of him published in the *Times* during the outing shows a youthful Curtis with regular features and a full head of hair, his glasses contributing to a studious, slightly owlish look. An accompanying letter from a member of the climbing party, though sent prior to the failed assault on Olympus, proclaimed Curtis to be the "hero" of the campaign. The correspondent seemed most impressed by the venture's precise organization and the six-course dinners with two desserts that allowed the Mountaineers to live "higher than they do at home."[10] When the first, larger group landed at Seattle a *Times* reporter wrote about its members who were "loud in their praise of the management of this expedition" and who assigned "great credit for its success to Curtis and his first assistant, L. A. Nelson." A later appraisal presented a sterner Curtis who, though a "round-faced, mild-looking man" was "an autocratic leader with an iron will, to which men almost twice his size meekly yielded—for the time at least."[11]

There was more to Curtis's leadership than making and enforcing tough decisions in the midst of snow, rain, plummeting morale, an injury, and almost nonexistent visibility. His careful planning ensured the advance understanding of group rules and roles, plus the smooth handling of all the necessary equipment and personal baggage. Curtis and his assistants, Nelson, Eaton, and the remarkable Montelius Price, organized every detail meticulously and innovatively. Two sets of comprehensive instructions, the second expanding and modifying the first, appeared in the March 1907 issue of *The Mountaineer*. Later thoughts concerning some original requirements, revised boat schedules from Seattle to Port Angeles, and other unanticipated circumstances required more changes.[12] The realities of the field forced quick, novel solutions. The base fee of $40.00 per person proved "utterly inadequate" to cover everyday expenses plus trail building, so Curtis, Nelson, and Price persuaded some Port Angeles residents to cover the cost of sixty miles of new trail. Finding the charges for packing "prohibitive," they bought their own pack train.[13]

As it was the club lost $304.56, at the time a considerable sum for a small, private organization. The financial report blamed the difficulty of estimating expenses in advance, and the fact that too many members signed up late for the trip. The board of directors voted to make up the deficit from the treasury. There may have been some carping about the outing committee's inability to control expenses, but in every other way the expedition was a success.[14]

The 1907 outing was not Curtis's first Olympic experience. Years later he reminisced about when he "first knew the range, in 1891," although his focus was not on where he went or what he did at age sixteen but rather on the changes since then. He recalled how the area in 1891 "was practically an unbroken wilderness but now [1915] the greater part of the range can be reached by some form of trail," while "more and more… people are coming to love the heights that can be reached only by mountain trails and I have watched with great interest the development of trails in the Olympics."[15]

Curtis's 1891 visit was significant in a way beyond his immediate concerns. The early 1890s saw vast tracts of peninsula timber saved in the public domain, the Olympic Forest Reserve, but transferred to private ownership within a decade by fair means and foul. The forests of far western Washington, once considered too remote for use, looked more attractive to railroads and lumber companies after they stripped other timbered regions of their wood. To be fair to powerful, determined men such as timber baron Frederick Weyerhaeuser and James J. Hill, the president of the Great Northern, the demand for lumber in their growing republic seemed, if uneven at times, relentless and unceasing. When the gloomy, rain soaked forests of the Olympic western slope came into view, the apparently cheapest and best way to harvest them was to cut them quickly and indiscriminately, taking the best species and trees, leaving the rest to rot on the ground.[16]

So the timber industry came to the Olympics with clear-cutting, denuding the landscape, encouraging erosion, and destroying the dark, forbidding thickets that supported a diversity of life, including the famous Roosevelt elk. Clear-cutting increased the danger of fire, a danger mitigated by about 150 inches of rain per year falling on the western slope of the Olympics. The mountains themselves caused the heavy rains when they blocked moisture-laden landward winds from the Pacific, forcing them to drop their burdens on the western slopes. Devastating fires occurred anyway, as President Franklin D. Roosevelt would discover some three decades later.

There were rational people in the timber industry who questioned clear-cutting, but not as much as they deplored the fluctuating prices and demand for their product. The problem of clear-cutting from their viewpoint was less the environmental damage it did and more its encouragement of unrestrained production during flush times. Timber overproduction and the overbuilding of mills during times of strong demand led to surpluses, falling prices, and severe competition when demand softened. This boom-and-bust cycle hampered efforts to balance supply and demand through careful planning for the future. Industry leaders and their trade associations were aware of the problem but lacked the power to enforce restraint on all producers during the good times. Slowly, too slowly for preservationists, clear-cutting fell out of favor. Its replacements were conceptions more likely to balance supply with demand: "selective cutting" and "sustained yield." Selective cutting felled only those trees mature enough to be processed profitably, leaving smaller trees to be harvested later. Selective cutting was more expensive because it required choosing individual trees, but it decreased the dangers of fire and erosion. Sustained yield involved cutting no more timber in any one year than could be replaced by matured trees in the next. Both practices were impeccable in theory but divisive in practice. Critics charged that lumbermen gave them no more than lip service.[17]

Curtis's reaction to all this was not the simple developmentalism that his detractors assumed. He was, depending on circumstances, a conservationist, a preservationist, a recreationist, and what would later be called an environmentalist. Where Washington State was concerned he was especially sensitive to maintaining and expanding a statewide economy dependent on timber, agriculture, tourism, trade, and, where feasible, mining. His interest in the state's welfare was a subset of his conservationism, his belief that all natural resources had to be carefully, rationally managed under federal supervision for the benefit of the people. Making conservation work on the Olympic Peninsula demanded thoughtful development because, as he wrote with its resource history in mind, the area was "so brainlessly exploited in the past."[18]

Curtis the preservationist ardently advocated saving scenic areas from destruction while opening them to recreation, especially in national parks, forests, and other federal lands. The headwaters of the Elwha River, he wrote in 1915, were "God's own country." At the time "God's own country" lay within the Mount Olympus National Monument, the core of which he had tried to preserve as a national park. His reverential language was hardly that of an exploiter. Nor was he arguing for turning "God's own country" into an easily accessed theme park. People who wished to experience it would have to hike in, just as his 1907 expedition did.[19]

On the other hand Curtis was no recreationist snob. He recognized that a modest number of national park or national forest visitors were concerned with long forays into the boondocks, as the previous chapter demonstrates. He advocated all types of recreation from the strenuous to the passive. Vigorous mountain climbing, horseback riding, hiking, and camping remained his personal preferences. Preferences aside, he encouraged road construction in the national parks for those who preferred to observe their wonders from the inside of a car. Then there were those who, having arrived by car or bus, wished only to loll about the porch of a park hotel. These forms of recreation were also legitimate in his view.

As far as the Olympic Peninsula went, he was too involved with Rainier to support, oppose, or worry much about continuing efforts to develop a park there. By no means did he ignore the peninsula or its mountains. In 1924 the Aberdeen Chamber of Commerce commissioned him to travel through the Olympic National Forest, photographing as he went. The strongest conviction emerging from the month-long venture was his belief in Forest Service control of all of the peninsula's public lands. That conviction required abolishing the Mount Olympus National Monument and the Interior Department rules prohibiting hunting, lumbering, and mining within its borders. Whether or not he decided his stance before or during the trip, he publicly announced it afterward.

On July 6 Curtis and his party arrived at the Quinault River, at what would become the southwestern corner of the future Olympic National Park. (The expedition would return on August 7.) The party comprised Curtis, Harold Olson, a "young, husky reporter" for the *Aberdeen Daily World,* and R. E. Voorhees, a Quinault Indian who served as packer, cook, and guide.[20] During the month the group traveled northward from the Quinault to the Sol Duc River, turned east to the Elwha, then south to the Quinault. Olson estimated that the trip covered one hundred miles by car, and 350 miles by foot, including the hooves of the three riding and two pack horses.[21] Curtis brought two cameras. An early Graflex Speed Graphic, a "press camera" using fast celluloid film, recorded animals, while a traditional box camera with glass plates and a tripod preserved scenic views. He snapped at least 230 photographs, many of them later converted to more than 150 colorized lantern slides for his lectures on the Oylmpics.[22]

Olson's stories in the *World* described in booster fashion the timber, scenic, mineral, agricultural, tourism, and game animal resources of the peninsula. The young reporter sprinkled his articles with asides illuminating Curtis's personality and attitudes. Early in the trip Curtis taught his companions pinochle, a card game similar to contract bridge, "which he surely regretted afterwards," whether because they regularly bested him or because they urged him to play cards too often, Olson did not explain.[23] Olson, though living nearby, had not tramped to the heart of the Olympics and doubted Curtis's descriptions of their alpine majesty. "His distinctive ability to make big stories out of little ones did not induce me towards greater gullibility." Once he saw the Olympic

Mountains, he believed. Their "awe-inspiring beauty" overwhelmed him.[24] When the party passed an evening in the comparative comfort of one of the scattered farms in the Hoh Valley, Curtis experienced a "rough night," unable to sleep well on the feather bed thoughtfully provided by his hosts. Finally "he rose and took his blankets out to the barn. Amid the bawling of calves and grunting of pigs he immediately fell into a sound sleep."[25]

An Olson article on game animals reflected the likely prevailing sentiment on the peninsula regarding the Roosevelt elk and other large species. Olson argued that all the animals once lived in a natural balance but man, meaning non-Indian humans, had upset this harmony. Therefore man must regulate animal numbers to restore the balance that his activity disrupted. Tourists loved the elk, thus the primary focus should be on their preservation. An imbalance among the elk themselves led to an oversupply of bulls, so they should be selectively killed. Cougar and bear, natural predators, were bad for elk, therefore they "must go." The forest supervisor, Rudo L. Fromme, was ready to work with the state game department for effective animal management, a backhanded admission that state game laws designed to prevent poaching were ineffective.[26]

Fromme joined the group for the last week of its journey. He and Curtis took several side trips to photograph scenes. On one of these jaunts they became lost in a rapidly developing fog. Fromme wrote about the incident an unknown number of years afterward, and his memory of the location of their misadventure cannot be verified, but there is no doubt of their being lost on dangerous ground and in peril of their lives. Then the fog lifted as quickly as it came. They found their way to camp, where Olson and Voorhees, concerned for their safety, were relieved to see them.[27]

In his account Fromme insisted that Curtis had come, among other reasons, to scout the area for a possible national park. Some of the people who financed the expedition wanted a park, knew of his devotion to Rainier, and expected him to report in favor of an Olympic park. Instead Fromme, in several "heated" discussions, convinced Curtis of the desirability of Forest Service management. Fromme credited Curtis with being "a good listener as well as a person who craved all the answers." When the discussions became too intense, Curtis would break the tension with "Oh hell; let's have another pinochle game to go to sleep on."

Fromme did not list the arguments he advanced, but if the talks occurred as he described them, they may be inferred from Curtis's article in the *Aberdeen World* series.[28]

Curtis insisted that the national monument should not be made a national park because its area was too rugged and its boundaries too arbitrarily drawn for effective administration. It should be abolished and folded into the national forest. The Forest Service should administer the entire federal reservation because it could protect recreational areas "without sacrificing the remaining tremendous resources of timber, minerals, water power and agricultural lands." If the expanse became a national park there would be no development and "no return on the advertising value of the park." What he meant by "advertising value" was not clear. Lastly, roads and trails could be built as easily in a national forest as in a national park.[29]

Curtis expressed himself forcefully on other subjects when he wrote a post-expedition article for the weekly *Argus.* He praised the agricultural bounty of the Olympics' western slope: at one "ranch" his party dug potatoes, "ate raspberries until we could eat no more," and walked through rye over their heads. The crops required only the peninsula loop road, still under construction, to bring them to market. He also revealed his love for trees. He and his companions rode and walked by "Western Red Cedars that were old when Columbus first saw the continent," and Douglas firs that "lift themselves 300 feet to reach the sunlight, their symmetrical columns free from limbs for half their height." It was "a pleasure" to leave the mountain peaks for the "deep forest valleys, to sit around the camp fire and watch the light play on the trunks of the giant trees." And there were the elk. Curtis crawled through brush with his Speed Graphic camera to draw close to a small herd, getting in five shots before the wind shifted and the elk smelled him, then three more while they fled. He snapped all the photos "within easy rifle range and still some think it sport to kill elk." Roads had to be built to bring tourists to the forests and the animals. "God hasten the day when the trails of the Olympics shall be the highways of the Olympics," a view he later modified.[30]

That was in 1924. Not until 1929 did a Herbert C. Hoover administration proposal force Curtis to again marshal his reasons against a national park for the Olympics. Then the ebullient, peripatetic Ernest Walker Sawyer, an Interior Department official charged with stimulating the western economy, asked Curtis for a memorandum on an Olympic park. Curtis replied that he had once favored such a park but had changed his "mind about the type of development best suited to this area." He did not reach the "principal reason" for his objection until the middle of his reply to Sawyer, when he declared that it would be almost impossible to draw administratively sound boundaries for a park, because including all of the mountain area would encompass the "river valleys which are not of special National Park character" but "contain timber which we must expect to use." A park boundary around the mountains and their approaches would also include "considerable mineral wealth," so for reasons of development the area "should remain a part of the National Forest."[31]

Taking a different line of attack, Curtis argued that there were "but two considerations in favor of a national park," expanded, improved tourist accommodations and the still undefined "advertising value" of recreation. On the other hand, there were three arguments against tourist and recreational development under the aegis of the National Park Service. The first was unwelcome competition with Rainier. All the "funds which we can reasonably hope to secure" would be needed for Rainier. That park's administration required "many millions of dollars" to complete a road system of only limited value until it was finished. Recreation would "be best served by concentrating on Rainier until development is complete." Besides, if the Olympics became a park, so should "a similar region around Mt. Baker and Mt. Shuksan in the northern part of our State, also Mt. Adams and Mt. St. Helens," obviously too large an order for the federal government to fill. Having raised the specter of rampant national park growth, Curtis banished it by an indirect reference to the Forest Service's Cleator Plan of 1927-1928 for roads, trails, and areas reserved from timber harvest in the Olympic National Forest. He also pointed to the service's Primitive Area Regulations of 1929, providing for recreational uses in all national forests.[32]

Whether creating recreational reserves in "primitive" areas was a Forest Service ploy to defuse public opposition to timber cutting on federal land, an effort to compete with the Park Service in the recreational realm, a genuine attempt to provide adequate facilities for leisure activities in the national forests, or some of all three is open to debate. The point here is Curtis's belief that the Forest Service recognized "the public's right to the use of the forest for recreational purposes" and therefore the service should "go ahead with their development plans rather than to materially increase the number and area of our National Parks." He then presented his final argument against a park on the peninsula, that a park of reasonable size would include second-rate scenic areas. Only the best natural environments should be saved for inclusion in national parks.[33] He often returned to this theme.

By the end of the 1920s Curtis could assume that he had deflected the Hoover administration's interest in a national park. The stock market crash of October 1929 and the intensifying focus on what became the Great Depression dimmed the outlook for a national park in the faraway Olympic Peninsula. Whether because of Fromme's entreaties, the Cleator Plan, or because of his own convictions, Curtis had decided in favor of Forest Service control of the future of the Olympics. Summarized, his decision developed from his overriding belief in conservation, that is, the intelligent use of the natural resources of Washington State under federal control. Next in importance came preservation, which the Forest Service had to commit to if it wished to advance its third stated objective, recreation in the national forests. There could be no satisfactory recreation in a clear-cut moonscape. Curtis was ready to defend his position on these bases, and would have the opportunity to do so soon enough.

CHAPTER TEN

The Olympics: Battling Against the Olympic National Park

THE GREAT DEPRESSION revolutionized the political landscape. The resurgent, victorious Democratic Party would prove more congenial to a large Olympic park, partly because the incoming Roosevelt administration was dedicated to the preservation mandated in the national park act of 1916. Park issues were, however, but one dimension of Republican Curtis's aversion to Franklin D. Roosevelt. His reference to Roosevelt as "the boy wonder from New York" nicely encapsulated his attitude toward the new president's aristocratic eastern antecedents.[1] If Curtis mistrusted Roosevelt's preservationist enthusiasm and sympathy with the National Park Service, the new president confirmed his suspicion in June 1933 when, by executive order, he transferred all national monuments to the Park Service.[2]

By the early 1930s other Curtis attitudes had hardened into resolve. As his derogation of Roosevelt suggests, he was an unabashed provincial who was skeptical of easterners and their intentions concerning western lands. He was unsympathetic with the wilderness purists and landscape mystics of any geographic location who clamored for vast areas of primeval acreage. There were few of them, compared to those who liked their scenery taken in more civilized draughts. Rainier was highly developed, yet only 1 or 2 percent of it had yielded to roads, tourist lodgings, or other artifacts of an urbanized culture. The other 98 percent or so of Rainier "alone will accommodate all the people I know whose habits are such that they cannot enjoy a mountain if someone else is looking at it."[3]

Concerning the Olympic monument, the Park Service was unenthusiastic about taking it over, and there were good reasons for its hesitation. The monument, at about 330,000 acres, featured rows of peaks like dragon's teeth, with few of Rainier's gentle lowlands and broad upland meadows. Its rough terrain imposed serious administrative burdens.

The Olympic National Forest surrounded it, making any improvements dependent on cooperation with the Forest Service. The only feasible solution was to convert the monument into a national park, taking some Olympic National Forest territory with it to preserve forests and other attractions, create a coherent recreation area, conserve wildlife, and prevent exploitation such as mining. It was a situation fraught with practical and political uncertainty, but Roosevelt's Secretary of the Interior, Harold L. Ickes, would not countenance any hesitation. A preservationist, Ickes was also a ruthless administrator, a tough bureaucratic infighter, and a relentless expander of his Interior Department domain. He pushed the Park Service in the direction it had to go once the monument became its responsibility.[4]

The history of the battle for and against the park has been recounted several times. It is enough to note that the leading environmentalist group, the tiny Emergency Conservation Committee (ECC), emerged with a park of the largest possible size. It did so not primarily because two of its members, Rosalie Edge and Irving Brant, had Secretary Ickes's ear. Nor did it succeed because of its slanted publicity, arrogance, self-righteousness, intransigence, and bitter condemnation of any organization that deviated from its program. The committee shared those traits with others in the fight. The advocates of a huge national park won because Roosevelt favored a large park, because a large park appealed to Ickes's preservationist sympathies and empire-building proclivities, and because the Park Service of the Interior Department seized an opportunity to aggrandize itself at the expense of the Forest Service, still situated in the Department of Agriculture, despite Ickes's urgings to transfer it to Interior.

Just what was the ECC? A look at the group's members and its activities reveals its origins and methods. The godfather of the committee, Willard Gibbs Van Name, was a first-rate zoologist at the American Museum of Natural History in New York. He was also a preservationist zealot who stimulated Rosalie Edge's interest in saving birds and forests from exploitation. His zeal left little room for self-indulgence. He lived alone. An ascetic, abrasive misanthrope, he lived on a modest salary and inheritance and spent his time on written pleas for preservation combined with denunciations of those who cloaked their subservience to sportsmen, arms manufacturers, and lumbermen with fine phrases about saving trees and birds. Van Name's forebears included the distinguished biblical scholar J. Willard Gibbs and the astonishingly brilliant mathematician, physicist, and chemist Josiah Willard Gibbs. His unusual befriending and tutoring of Edge led directly to the committee's founding.[5]

Rosalie Edge enjoyed a lineage as distinguished as Van Name's. She was a member of the patrician Barrow family of New York, and married the prosperous railroad and bridge engineer, and, later, stockbroker, Charles Noel Edge. Her evolving views on conservation stimulated by Van Name and others, she fought the National Association of Audubon Societies' manipulation of its wildlife sanctuary for the benefit of hunters and trappers. Almost single-handedly she established the Hawk Mountain Sanctuary in Pennsylvania. She created the ECC partly to provide a cover for Van Name's anonymous, provocative writings on conservation and preservation.[6]

While she grew in stature in the preservation movement Edge suffered serious personal reverses. She became estranged from her husband, her daughter, and eventually from Van Name and others. Indifferent to the effects of her caustic attacks, she could be emotionally fragile when counterattacked. The Great Depression forced her into reduced circumstances. Reduced circumstances for Rosalie Edge were a four-room apartment filled with antique furniture, Oriental carpets, James McNeill Whistler etchings, and leather-bound books. Her living room commanded a view of New York's Central Park. From there her letters, pamphlets, and editorial writings poured forth to fill the mailboxes of thousands of preservationists.

The eastern origins and preservationist stridency of Van Name and Edge alienated Curtis. The background of the third committee member, Irving Brant, was closer to Curtis's, though that did not endear him to the Seattle photographer. Brant never endured anything like a penurious existence in the upper Midwest or Port Orchard. He was a talented writer and editor whose birth into an Iowa newspaper family gave him a leg up in the business. His formidable six-volume study of James Madison established him as a scrupulous historian, though he found his true calling in conservation advocacy. He wrote, under Edge's name, the principal committee pamphlet on the Olympic park issue, among many other tracts. He enjoyed close ties to Roosevelt and Ickes, becoming a "personal assistant to Secretary Ickes and speech writer for President Roosevelt" on the federal payroll.[7]

Other actions and sentiments worked in favor of the park idea. Curtis and others had tried for a park in the past, leaving a favorable if unquantifiable feeling for a park. Some Forest Service officials were ambivalent about maintaining a national forest merely for resource exploitation, and were anxious to present their organization as being just as committed to preservation and recreation as the Park Service. Too, Secretary of Agriculture Henry A. Wallace gave only modest support to the Forest Service in its fight against the park. The logging industry opposed preserving large areas of merchantable timber, but it was part of a business community held in low esteem in the 1930s. There is no impartial evidence of a shift in attitudes of Olympic Peninsula residents, but they appear to have moved toward greater acceptance of a park. The only, if critical, point of contention was the size of the park and thus the volume of timber and other resources removed from exploitation. As early as 1916 a Forest Service official declared that he had "no doubt of a decided local sentiment in favor of a National Park in the Olympic Mountains." Curtis recognized the threat to Forest Service control in 1924, when he urged developing recreation sites in the national monument to "forestall" a "demand" for converting the monument into a national park.[8]

Ten years later, however, Curtis had come around to his original position in favor of a national park. A compromise in the Seattle Chamber of Commerce, of which he was the guiding spirit, recommended converting the existing monument into a park, while excluding the rest of the land that had been part of the original monument. "I have just read Curtis' recommendations," one Park Service official wrote to another in September 1934. "I say they are not bad. He really favors a park! What changed him?"[9] What changed Curtis was his realization that practically all influential people on the Olympic Peninsula wanted a park of some sort. Curtis had not altered his opinions about Roosevelt, the New Deal, the eastern advocates of a massive park, or the desirability of Forest Service control. Given the shift in local opinion, opposition to a national park in the Olympics did not make sense, thus his effort to balance resource conservation, recreation, preservation, and the survival of an Olympic Peninsula economy based on timber, water power, and mineral development. Further, he no longer feared competition for federal dollars between his beloved Rainier and a park in the Olympics. By the summer of 1934, the large appropriations of the late 1920s and the New Deal's public works spending had brought Rainier within sight of its full development by the standards of the day.[10]

Following a report from a Park Service field investigating team that suggested a park of modest size, the ECC intervened on behalf of a larger park than the report recommended. At the committee's behest, Congressman Monrad (Mon) C. Wallgren of Everett introduced a bill for the Interior Department in March 1935. Wallgren's bill would create a park from the existing monument and add 400,000 acres. Curtis objected to the measure as soon as he received a copy. The park, he wrote, would include ordinary trees found in other areas of Washington, thus the timber was "not of national park status." The bill would not adequately protect the Olympic elk. It would cancel the careful plans of the Forest Service to use timber on a sustained yield basis. It would reduce the interest in state-owned timber adjacent to the park because, with the park timber held inviolate, harvesting trees on state land only would result in higher costs. It would radically reduce the federal timber payments to peninsula counties. If the proposed park remained in the "primitive" state provided for in the bill, it would "be of little value from a recreational standpoint," because in his experience tourists were not "prepared to make long hiking trips into remote regions." The tentative boundaries were the work of easterners who had little conception of the realities of the Olympic area.[11] In fact, the reduced monument with no additions "constitutes a wilderness area large enough to take all the hikers in the United States and give them each plenty of ground to get lost in."[12]

Fear for the economic future of the Olympic Peninsula if logging were curtailed was implicit in Curtis's letters, and one of them contained a mild suggestion of widespread opposition to a large park.[13] But Curtis was explicit about serious opposition to the park as proposed in letters to his friend Cammerer, director of the National Park Service. In a letter of April 27, 1935, Curtis noted "hostility" to a large park, and in a September 12 letter he declared that unless the boundaries were shifted "our state will endeavor to block the passage of the bill."[14] Curtis's assessment of state public opinion was doubtful, but there was, as Wallgren asserted in a letter to Curtis, little chance of the bill being voted on in 1935.[15] Nineteen thirty-six would be different, probably, so Curtis prepared an article stating his position in full with the hope of influencing the future of the Wallgren bill.

His effort appeared in the April 1936 issue of *American Forests and Forest Life* as "The Proposed Mount Olympus National Park." He reiterated themes introduced in his private correspondence: the park was unpopular; the Forest Service would manage the timber well; counties would lose income; the state's forest lands would be less attractive to loggers; the elk would not be well protected; there was enough recreational development in the national forest and the national monument to satisfy all who wished to venture into the wilds; and the timber in the proposed expansion area was second-rate, a clear violation of the standards for national parks set with the founding of the Park Service. He extended and strengthened those arguments, but also stressed the need to maintain the economic prosperity and social vitality of the Olympic Peninsula. Because timber was its principal crop, it had to be available, not locked away in a national park. The "estimated" sustained yield cut in the national forest "would support a wage earning population of 27,520." Locking up the trees proposed for the national park would reduce the "supported population" to 10,870.[16] If the trees could be harvested rationally, then there could be "permanent communities of trained forest and mill workers." There would be some "farm homes" nearby, part of "a population of many thousands of self-supporting, industrious and happy people with a livelihood guaranteed for generations."[17]

John B. Yeon, a young Portland architect, challenged Curtis in the June issue of *American Forests*. The magazine's editors believed Yeon's piece to be "the clearest and strongest argument yet made for the park." As did most but not all park supporters, Yeon avoided detailed discussion of the human and economic losses involved in creating it. He praised the Wallgren bill for considering "the preservation of the area's natural geographic and biotic features," thus saving the "scenic, recreational, educational, and inspirational resources of the region." The bill, which he wrongly imagined was a Park Service product, would thus preserve "a world enchanting in its detail, thrilling in its vastness, and with a prevailing power of magic to heighten the spirit and imagination of the visitor." Yeon coupled his landscape mysticism with practical considerations. He found that the touted methods of forest preservation, such as selective cutting and mixed use, destroyed landscape values. Big trees were no longer needed for lumber because "fabricated timber made up of many small laminated pieces" did a better job of spanning large spaces. The Forest

Service's primitive areas were a sham, first, because they were mere strips surrounding natural or man-made features, and second, because they would be sacrificed once other timber was taken. Yeon did not explain how his forecast of forest doom dovetailed with his conviction that the lumber market was "glutted," but his article proved that *American Forests* treated both sides of the park issue fairly.[18]

To return to Curtis, he illustrated his own *American Forests* article with three rhapsodic photographic views of the Olympics, as he did his testimony before the Committee on Public Lands of the House of Representatives. The committee took testimony on the Wallgren bill for nine days, beginning the same month as Curtis's article appeared. Curtis, speaking as the representative of the Aberdeen Chamber of Commerce, reiterated most of the points in his article. He gave ground to the proponents of a large park when he agreed that a small number of large trees west of the national monument could be included in the park. He argued for including the Olympic Mountains visible from Puget Sound, a long-standing enthusiasm of his. Beyond these two alterations, he refused to budge. The core of his testimony opposed the bill because it would "lock up" useful timber in a national park, thus eroding the base of "a civilization of people that were producing something that is marketable."[19]

The word *extensive* is a barely adequate descriptor of the hearings. They proved to the committee that the Wallgren bill was nothing if not controversial, as the more than five hundred printed pages of testimony and exhibits revealed diametrically opposed views. Depending upon the witness, sustained yield—cutting no more timber than yearly growth—was feasible or unfeasible; peninsula tourism had great potential or almost none; the timber industry was vital or staggering on its last legs; there was more than enough timber on the peninsula or it was rapidly disappearing; the Forest Service was best equipped to handle the federal property involved or the Park Service could do a better job; the economy of the peninsula would be devastated if the bill passed or the fate of the peninsula was a national matter transcending any local impact. The committee reported the bill to the House with a positive recommendation and without amendment, but it did not pass before Congress adjourned for the summer.[20]

Curtis's article and the hearings produced two major outcomes, neither good for him. His article led directly to the death of the Rainier

National Park Advisory Board and the end of his involvement with Rainier. His pro-development testimony aroused opposition of another kind. "Whether or not I can ever regain my reputation in the State is an open question," he wrote after his return from Washington, DC. "Nearly everyone I have seen since wanted to know how come I had joined the lumber barons and the predatory interests."[21]

The charges had little merit. Curtis was hardly the tool of grasping lumbermen, for he declared in his testimony that the function of timber cutting and lumber mills was to provide jobs. The lumbermen would have to market their product and make a profit while doing so, of course. Nowhere in his testimony, his writing, or his surviving correspondence, however, did he express any concern for mill owners or lumber companies. If the proposed park were created, lumbermen, in his opinion, would operate strictly on self-interest, closing the mills if too many trees were preserved in the park, or paying less for state timber set aside for support of the schools if they could not combine it with a large volume of national forest or privately owned timber. So far as he was concerned, the primary function of mill owners was to pay the salaries and wages necessary to a stable, secure American way of life on the Olympic Peninsula.[22]

Part of Curtis's testimony could be faulted because of his disingenuous statement in the 1936 hearings that he "believed in the creation of a national park in the Olympic Mountains for many years." He had supported a park before, but ended his advocacy in 1915 and did not accept a park again until 1934, when a park of some size seemed to be a certainty. His assertion that the Park Service refused adjustments in the boundaries of the Wallgren bill was not entirely correct either. The boundaries, he thought, were "only a tentative matter," adding, "I felt, until I came here, or shortly before that, that we were going to be given an opportunity to discuss these boundaries, and fix them."[23] In support of his contention he introduced three letters during the hearings. One from Wallgren, dated April 16, 1935, declared his bill to be "only a starting point for legislation," but that letter was a year old, written before the Interior Department and the Park Service built momentum on behalf of Wallgren's proposal.[24]

The other letters, dated May 4, 1935, and November 20, 1935, were from Cammerer. In the May letter the parks director admitted that "doubtless some revisions of the boundaries" in the Wallgren bill could be made, but in the November letter he was willing to consider only an eastward extension of the boundary to include the Olympics visible from Puget Sound, an increase, not a reduction, in the Wallgren limits. Moreover, Curtis omitted Cammerer's October 23 letter in which the parks administrator refused to recede from including the timber west of the monument, trees that Curtis wanted left for cutting. Cammerer could hardly do otherwise, for his boss, the irascible Ickes, supported the bill as it was. Cammerer's mild flexibility on boundary changes was some justification, however modest, for Curtis's continuing effort to have much of the Douglas fir, spruce, and hemlock west of the monument excised from this and subsequent legislative proposals.[25]

Curtis also objected to what he believed was the cut-and-dried tone of the hearings. The Park Service, he thought, had stacked the deck. "Judging...by the wording of some of the editorials which were submitted as evidence," he wrote to Washington's Governor Clarence D. Martin, "one cannot escape the conviction that the National Park Service had prepared the editorials and thoroughly canvassed the committee before the hearings were held." The result was that "the committee members had, to a large degree, already decided the question before the hearings."[26] "The case appeared to be settled before we got there," he complained to another correspondent. "Thus the hearings were but a little side show put on for the benefit of somebody, I don't know who."[27]

Curtis was absolutely wrong about the committee having made up its mind beforehand, as a fair reading of the hearings demonstrates. It is true that editorials, resolutions, and other material in favor of the Wallgren bill went into the record, but so did statements in opposition. The committee allowed Curtis to include the Wallgren and Cammerer letters, and reports from the Park Service investigating team, in support of his contention that the park was too large. The committee members consulted photographs, pored over maps and charts, and asked pointed questions of witnesses. "I believe the gentleman knows what he is talking about," a committee member said of Curtis before asking a question. Curtis's dismissal of the hearings was unjustified, considering the sheer volume of testimony and exhibits, and the time and attention given to the issues.[28]

The committee's reporting the bill favorably and without amendment could be construed as lending credence to Curtis's claims of a prior decision, but there is no evidence tying the report to a preexisting committee attitude.[29] Nineteen thirty-six was an election year, and Congress was anxious to wind up its business and get on with campaigning. If the public lands committee was concerned with hustling the bill around a legislative logjam, it would have better served its purpose by holding perfunctory hearings for a day or two, then moving the bill along, rather than holding it up for nine days of testimony. The greater probability is that the members approved it without amendment because they knew that it was unlikely to be voted on before the House adjourned, to say nothing of its passage through the Senate. Curtis may have been closer to the truth when he asserted that some committee members would not want to vote against Ickes, "as he has the final say in the allocation of vast sums of money," both as secretary of the interior and head of the Public Works Administration.[30]

Curtis continued to hold the Park Service and its home department responsible for what he believed was only a show. His conviction justified him in his continued attacks on the Park Service for wishing to preserve what was legitimate commercial timber. Another principal argument was that, as he wrote Cammerer in June 1936, "the sentiment of the people of Washington will, I believe, be almost unanimous in opposition to any national park." Curtis advanced this assertion to defend his plan for a new committee study of the Olympic park question. He offered no evidence for statewide opposition to an Olympic park and overlooked popular support on the peninsula for a park of some size. Cammerer, tired of Curtis's importunities, rejected his call for another committee study and by implication his threat of popular disapproval. He replied that during the recent hearings, "the pros and cons of the situation were given thorough consideration. To have another committee go over the same questions would be duplication."[31]

As noted, Wallgren's first bill succumbed to the 1936 adjournment rush. After his overwhelming reelection, Wallgren returned in 1937 with a revised bill extending the park's boundaries to the east, to the mountains visible from Seattle and other Puget Sound locations. Among other changes, the new bill reduced the Douglas fir, spruce, and hemlock takings to the west. Wallgren made these changes in response to the 1936 hearings, which demonstrated that his first bill was controversial, and to a Washington State Planning Council (WSPC) report proposing a smaller park. The council, "created during the depression emergency to coordinate state and federal economic programs," was concerned with the vitality of the Olympic Peninsula's lumber economy. The Park Service accepted the new Wallgren bill partly because it appeased the locals and partly because the Forest Service would find it difficult to oppose a measure that returned some 138,000 acres to its domain. The revised bill's surrender of those acres enraged the ECC, which did not have to worry about local opinion, the local economy, administrative problems, or interservice rivalries.[32]

In a letter to Cammerer, Curtis denounced the new bill for the reason that it still contained too much land to the west "not of National Park caliber." By 1937 he was, however, bending to reality, telling Cammerer that "the state will bitterly resent the passage of this act."[33] He no longer pretended that public opinion could block it. In the end neither Curtis and his allies nor the people he called "the fanatics in the east" would control the size of the park.[34] President Roosevelt, more than anyone else, would determine its size. He handed the "fanatics" a victory when he insisted on a park even larger than the one proposed in the first Wallgren bill. His decision came after he visited the area on September 30 and October 1, at Ickes's suggestion, and as an extension of a previously planned western trip.[35]

Roosevelt was able to push the park through Congress for two reasons. The first was his firm grip on the public's imagination. The luster of his 1936 electoral triumph had somewhat dimmed during his unsuccessful effort to expand the membership of the Supreme Court and the beginning of the recession of 1937–1938, but he remained enormously popular and therefore extremely powerful. The second was his enthusiasm for a park. He was so impatient when the first Wallgren bill stalled in Congress that he wanted to transfer the added 400,000 acres proposed in the bill to the existing national monument by executive order. With the transfer, the vast addition to the monument would become the domain of the Park Service, and the only action left for Congress would be to turn the expanded monument into a national park. His attorney general advised against such a bold move, so Roosevelt dropped the idea, but he still wanted a big park.[36]

Roosevelt's trip proved fatal for those who, like Curtis, opposed a large park. Two deceptions bracketed the president's stay on the Olympic Peninsula, both serving to solidify his commitment to a sizeable reserve.

One triumphed and the other backfired, but each in its way moved the president forward on the path he had already chosen. The failed deception, a Forest Service blunder, will be described later. The first, successful, bit of guile is enshrined in environmentalist mythology as an ardent youthful plea to Roosevelt for a national park. At least three books present the story of a children's crusade that touched the heart of the magnanimous president, who responded in avuncular tones to the youngsters' yearning for scenic preservation.[37]

As with many myths, a dash of reality conferred plausibility. The story unfolded after Roosevelt arrived by destroyer at Port Angeles. His motorcade passed though the town, stopping briefly at the Clallam County Courthouse, where he saw a banner proclaiming: "Please, Mr. President, we children need your help. Give us an Olympic National Park." School children had nothing to do with the banner. As one participant later admitted, it was concocted at a house party of Port Angeles business people who favored a national park, though not one as large as the park became. Knowing that Roosevelt's motorcade was to stop at the courthouse, they hung the banner in front of the building. Their ruse fooled the president. Roosevelt thought that the severely classical courthouse was a school and that the children who turned out to greet him were its pupils. He wove his extemporaneous speech around the banner, calling it "the appealingest appeal that I have seen in all my travels. I am inclined to think it counts more to have the children want that park than all the rest of us put together." He told the "boys and girls" that they could "count on my help in getting that national park."[38]

Following dinner at a resort at Lake Crescent, west of Port Angeles, Roosevelt held court in a cabin. Local and regional Forest Service officials had maneuvered to play host to the Olympic Peninsula portion of Roosevelt's visit, but many other people were present also. Roosevelt displayed his partiality for the Park Service when he insisted on visiting with park rangers, thus Tomlinson and Preston P. Macy, custodian of the Mount Olympus National Monument, were included in the presidential entourage. They wrote extensive memoranda preserving Roosevelt's comments, the atmosphere of the postprandial meeting as proponents and opponents of the park came and went, and of events during the next-day auto tour of the national forest. During the course of the evening, Roosevelt made it clear that he wanted a big park, the bigger the better.

Among other comments, the president dismissed the idea of water power in the park area, deprecated the belief in the production there of manganese at a reasonable cost, and declared that the second Wallgren bill did not include enough park acreage. He believed that the timber reserved in the proposed park was "more valuable for its recreational use than for lumber."[39] Recitations of logging volume, past or future, or payrolls lost, had little meaning for him. The important thing was to make the park large enough for future generations.

Roosevelt's disinterest in the survival of the peninsula's traditional timber industry mixed with agricultural pursuits is surprising at first glance. A visionary conception of family farming combined with the farm breadwinner's part-time work in safe, clean, decentralized industry next door to the farm captivated such diverse characters as Henry Ford, Frank Lloyd Wright, and Roosevelt himself. Part-time farming, combined with hunting, trapping, fishing, and part-time lumbering on the peninsula was not as arcadian as the proponents of marrying farm and factory would have liked, but it was real enough. Even if Irving Brant's pessimistic estimate of two and a half to five more years for the lumber industry's survival were true, which it was not, the industry's continuing as before would have breathed a little more life into the peninsula's economy. Unfortunately for the existing economy, Roosevelt, not to mention Ford and Wright, discovered that the farm-and-factory aspect of the back to the land movement wasn't easily doable.[40] But a big national park in the peninsula was doable.

Roosevelt displayed some interest in one of Curtis's hobbies, a national park of high scenic standards. "We must have more large timber," he declared. "Eastern people enjoy the big timber and especially the hemlock. Extensive areas of the large timber are necessary. Otherwise you can't have a unique national park."[41] He mentioned a large hemlock on his estate "which people come for miles to admire," but other than his reference to hemlock, he spoke about saving the largest area possible, against which goal everything else was a trifle.[42] Throughout the discussion Roosevelt displayed a remarkable knowledge of national park history, conservation measures such as sustained yield, the specifics of the two Wallgren bills, and the proposal from the WSPC.[43]

The next morning the motorcade continued through the Olympic National Forest while Roosevelt asked "many technical questions" about the national forest and the national monument.[44] The second deception, which hardened him against the Forest Service, occurred on this leg of the trip. Among the many signs that the Forest Service erected to guide the president was a large one reading Farewell, suggesting that his caravan was leaving the national forest. In truth the forest boundary was about two and a half miles farther. Some two miles of the intervening distance included a devastated stretch of clear-cut and burned-over timber. "I hope the son-of-a-bitch who logged that is roasting in hell!" Roosevelt snapped as he was driven through the scarred landscape. A small sign marking the real forest boundary either was removed or went unnoticed. When Roosevelt discovered the deception he was furious and ordered the transfer of the Forest Service official he held responsible for the trickery.[45]

In February 1938 Roosevelt called a meeting at which he told Wallgren and several others to develop a large park. Wallgren introduced the result on March 28, substituting it for his second, 1937 bill. The new proposed park was much larger than either of its predecessors and set off elaborate legislative and administrative maneuvering. Though the outcome was uncertain in its details, there was no doubt that any park emerging from the wrangles would be large. The Forest Service surrendered at a meeting held to develop the new park boundaries, leaving Forest Service employees who fought the park at the local level twisting in the wind. Secretary of Agriculture Wallace ended his tepid opposition before Wallgren introduced his third bill. Opponents of the huge park were by then fighting a rearguard action.[46] Curtis advocated a smaller park for all the usual reasons in a statement included in the perfunctory hearings on the third Wallgren bill.[47] Few cared.

By that time the focus was on the intricate political dance preceding the final passage of the third Wallgren bill. Significantly, Brant figured prominently in the negotiations. The legislation that Roosevelt signed on June 29 created a park of 648,000 acres, the same size as proposed in the second Wallgren bill, but allowed the president to expand the park to a total of 898,292 acres. Ickes dispatched Brant to write a report on the acreage to be included. After negotiations with state officials, Forest Service employees, and other interested parties, most of the alternate acres were folded into the park. Roosevelt proclaimed the additions on January 2, 1940. They were not enough to satisfy the president and Ickes, who wanted to preserve land along the Queets River, southwest of the park, as well as coastal land to the west. They could not take all they wanted under the park act, so Ickes began acquiring the land as a Public Works Administration project.[48]

Curtis remained an unreconstructed opponent of the Olympic National Park. His old friend Tomlinson and Arthur E. Demaray, the acting park service director, nevertheless requested his opinion about development of the new park. Curtis complied with suggestions for trails and other improvements, but learned that there was no enthusiasm in the Park Service for any more conferences about the park's size.[49] Tomlinson and Demaray were wise to ask Curtis's advice, for he knew parts of the area thoroughly. If they also intended a reconciliation, their overture failed, for he remained estranged from the Park Service. By then he no longer cared to carry on the fight, because he believed that people were tired of his discourses on park issues. It was "easier" for the public to believe "phrases coined by paid writers of the East."[50]

Curtis continued to speak his mind in his private correspondence. He could not contain himself after he learned of Roosevelt's proclamation and of the Public Works Administration's (PWA) intent to create the Queets Corridor to link the inland and coastal parks. "The Olympic Park reminds me of the ancient mustard plaster. Its purpose was a counter irritant. It made you feel so bad in a new place that you forgot the original trouble." He likened the Queets Corridor to the Polish Corridor, an outlet to the Baltic Sea arranged for Poland after World War I, an outlet obliterated by Germany at the onset of World War II. The idea of a corridor was "borrowed from the language of the Old World and I do not like any of the things we got from over there."[51] He did not object to the Pacific Coast park itself. In 1924 he advocated a similar park, although under state auspices in connection with the peninsula loop road's construction between the (then) national forest and the beach.[52]

The federal government's purchase of farms and other property in the Queets Corridor enraged Curtis. The Park Service was "driving settlers off" lands adjacent to the Olympic park, and worse, he was "amazed" at the indifference throughout Washington State over the loss of private land. "It is a modern Arcadia [sic] needing only a Longfellow to shock the world with its brutality." His ire centered on Ickes, who was attempting "a dictatorship by fear."[53] The only solution was to drive out Roosevelt's New Deal and return Ickes "to Chicago, stripped of all his power and let him try to make a living selling peanuts on a street corner."[54] One historian of the Olympic park saga dismissed those words as a "fantasy," but there was some hope for a Republican victory in 1940.[55] The Democrats triumphed, followed by a routine, post-election letter of resignation from Ickes to Roosevelt. Curtis sent a telegram to the president urging him to accept the resignation, which Roosevelt did not.[56] Curtis's death a few months later left his hopes for a reduction in the boundaries of the Olympic National Park unrealized.[57]

Nor did Curtis live to see the development of the Olympic park as a mostly roadless area where trails predominate. Today the Hurricane Ridge Visitor Center in the northern part of the park is accessible by car. So are a few other peripheral areas. Curtis would not have agreed with such severe limits on roads, although by the time the Park Service took over the area he had given up his 1924 notion of converting all the trails in the former national forest to automobile access. In 1933, before much of the forest became the national park, he advocated two routes then under consideration, both having a terminus at Lake Quinault, one bisecting the range and emerging north of the Elwha River, the other running east and north to the Hood Canal town of Brinnon. "If both these roads were built there would remain a vast wilderness," he emphasized. Branch or stub roads from the one or two through roads would encourage more people to visit the Olympics. He was not arguing for the projected road along the Elwha, writing only that the roads could have little impact on so huge an area.[58] Indeed Curtis opposed a Forest Service attempt to build a road south beside the Elwha, and the interest in the town of Port Angeles in its construction. The Forest Service abandoned the attempt and eventually built the road elsewhere. A correspondent thanked him for condemning the road, "this abortion…on behalf of… the fraternity who really enjoy that country as God made it."[59] Today no traverse roads run where Curtis suggested they could, although entrance roads and through trails mark both routes.

Curtis nevertheless underestimated the growing lure of wilderness to an increasingly urbanized, suburbanized population with easy mobility and rising income. "I find that there is very great increase in the interest in wilderness areas," he wrote in 1937, but from the context of his remarks he was thinking of those hardy types who were willing to ride horseback over rough trails. Probably he thought that any Olympic park would not draw many more people than the 5,880 who visited the national monument that year.[60] Private hoteliers agreed with him. They thought that a park "equipped with only campground facilities and trails" could not "attract more than a few hundred visitors," Elmo R. Richardson noted in his excellent review of the Olympic controversy. Yet, Richardson found, "75,000 came during the first season of 1938, a third of whom stayed for trips along the trails into wilderness areas." People kept coming to the park despite World War II, doubling their numbers by war's end. Visits soared in the postwar era.[61]

Curtis was also mistaken about the demise of the Olympic Peninsula economy. The beginnings of national rearmament in 1939, the boom years of World War II, the survival of some lumbering despite the vast park, all strengthened it. From the perspective of the 1930s, however, his worries were reasonable. In the midst of the catastrophic Great Depression it was asking too much of any but the most optimistic to envision powerful, relatively inexpensive cars, broad interstate highways built for high speeds, cheap fuel, passenger jet travel, and amazing increases in personal wealth and leisure. These dramatic developments shrank the world and collapsed the distance between the once-remote Olympic Peninsula and the rest of the country. Combined with the limitations of urban and suburban living, the improving access to the Olympics piqued an interest in wilderness almost unimaginable in the Franklin D. Roosevelt era.

Camping changed, too, in unforeseeable ways. As Paul Sutter explained in *Driven Wild,* campers' gear underwent a revolution as profound as any in transportation, income, leisure, or perceptions of wilderness value. As Sutter noted, urban-based technology invaded the wilderness with equipment that made camping, if not exactly easy, much less burdensome than before. Lightweight, warm clothing made of artificial materials, light but warm sleeping bags, and ultralight but stable tents, transformed the wilderness experience.[62] Sutter's list could be expanded to include compact, efficient stoves, freeze-dried or otherwise condensed food, and an assortment of electronic devices designed for communication or keeping track of one's whereabouts. Gone were the impediments of Curtis's time: heavy packs, blanket rolls, and sleeping bags, thick woolen clothing, unwieldy field kitchens, and ponderous tents bothersome to set up, secure, and repack. Only water could not be compressed or dispensed with, and it was abundantly available in the Olympic National Park.

In some other conclusions and contentions, Curtis was less than correct. The groundswell of public opinion against the park, posited in his letters to Cammerer, was purely imaginary. He had to be aware of peninsula opinion favoring a park, as he certainly was of the criticism against his testimony at the 1936 hearings, and of the statewide indifference to government purchases of private land after the park was established. He simply could not admit that these personally unpleasant outcomes meant that there was no significant public opposition to even a grandiose park. His unnecessarily censorious letters to Cammerer were milder replays of his attacks on officials during the Rainier name controversy, but unlike the battle of the previous decade, this time he had few allies. Those he had, a few politicians and bureaucrats, "the lumber barons and the predatory interests" and their chamber of commerce collaborators, failed to rally the people. His criticisms of the public lands committee and the Park Service were overwrought.

His eagerness to make the Olympics a showcase for Mather's ideal of including only first-rate timber and scenery in national parks was unjustified in absolute terms. Mather indeed upheld limiting national parks to only the most outstanding forest examples, but he compromised his convictions in individual cases.[63] Curtis's wholehearted embrace of the Forest Service because of its steps toward recreation and preservation was understandable. Mixing those two benefits with resource development was superior, in his view, to the Park Service's concerns for recreation and preservation only. Nevertheless, his enthusiasm for the Forest Service, given its long-term record, could have been more restrained.

Curtis's critics cannot claim exclusive possession of the truth, however. His love of the Olympic Mountains and their surroundings was deep and undiminished. No critic could claim to know the Olympics better than he. Curtis strode and rode horseback over trials, climbed mountains, camped, and took hundreds of masterfully composed, luminous photographs of his revered Olympics and their valleys. He would have scoffed at one critic's statement that he spent most of his adult life conspiring to destroy trees.[64] To leave his love for trees at one more example, he campaigned to bring the national park's boundaries to the western edge of Puget Sound, which, when done, saved thousands of trees.

Curtis correctly condemned the big park proponents when he assailed them for including great swaths of strictly commercial timber in the park. Brant admitted as much when he justified including "large areas of commercial timber, whose value would ordinarily be used as an argument for exclusion." The inclusion, Brant wrote, would save the rain forest and the winter range of the Roosevelt elk.[65] Curtis and Brant disagreed about the volume of commercial timber essential to preserve a semblance of the peninsula's original forest grandeur. The question of who was the more accurate was debated without agreement then and could be debated now, also without agreement. As for the elk, they were surviving without national park protection of their winter range. Curtis believed that he did not have to choose between the survival of the lumber industry and the survival of the elk, which he admired and wished to preserve.

What Curtis wanted to preserve on the peninsula, above all else, was a way of life based on lumbering and subsistence farming. His denunciation of the "dictatorship" of Ickes and the displacement of farms in the Queets Corridor proved how he agonized over removing a population rooted in the peninsula. Farming there was marginal, but that was part of its charm.[66] Peninsula agriculture embraced a Jeffersonian tradition: diversity of income, family unity, and the rejection of the superficialities of twentieth-century living. For him, that was enough. There was no suggestion of helping lumber barons by keeping people on the land.

In the end the advocates of a large park won. Curtis's eastern "fanatics," Ickes, the Mountaineers, peninsula residents who favored tourism and preservation, and the Park Service (although the latter would have settled for a smaller park) carried the day. Curtis, the Forest Service, and those who agreed with them lost. Yet Curtis, in losing, affirmed some enduring realities. A private-enterprise economy required successful private enterprise for its survival, and lumbering, however distasteful to some, provided jobs and income. Curtis loved the Olympics as much as any member of the ECC. His devotion was grounded in a more intimate acquaintance than any committee member could claim, or anyone else could, with the possible exception of a few forest rangers. Ickes was, if not a dictator, a lover of power whose empire building was, often enough, ruthless. Finally, there was the traditional, rural, mixed-income system on the peninsula, a system reminiscent of pioneering, one that Curtis fought to preserve. It is against these realities that the victory must be balanced.

A Curtis Miscellany

Among the many photographs of Native Americans in the Curtis collections is this 1902 study of hop pickers in the White River Valley south of Seattle. In this photo, posed for a payment, the women at the left are waiting for the pole puller to bring down the hop vine so that they may begin picking. For more on hop picking, see the book by Paige Raibmon in the bibliography. Curtis usually photographed Native Americans engaged in contemporary activities, in contrast to his brother Edward, who depicted them in pre-contact garb and settings. *University of Washington Libraries, Special Collections, A. Curtis 00438.*

Either Curtis or his erstwhile partner, William P. Romans, clicked the shutter at the launching of the lumber schooner Minnie A. Caine from the Moran shipyards on the afternoon of October 8, 1900. It is, however, typical of Curtis's best industrial work; lively, recorded at a moment of high interest, and incorporating humans to show scale. *Washington State Historical Society, Tacoma, 1943.42.180.*

A photograph that Curtis definitely did not take. He had no interest in the rituals of death, probably because he associated them with organized religion. When a client asked his studio to photograph a mortuary scene such as this one of a deceased mother and child dated November 11, 1920, he sent a member of his staff. *Washington State Historical Society, Tacoma, 1943.42.41106.*

A photographer for the *Spokane Spokesman-Review* probably snapped this photo of Curtis (facing camera) and Herbert C. Hoover, then Secretary of Commerce and later President, on August 19, 1926. Hoover was on tour, and Curtis was in eastern Washington to advance a "gravity" plan for irrigating more of eastern Washington. Curtis admired Hoover, although the "gravity" plan eventually lost to the "pumper" plan for an irrigation, flood control, and power dam at the Grand Coulee. Curtis cropped the original photo to eliminate others as much as possible. *Washington State Historical Society, Tacoma, 1943.42.2013.0.7*

In 1928 the Seattle Chamber of Commerce commissioned the Asahel Curtis Photo Co. to take a series of photographs of young women. The subject matter of "Miss Headly dressing" while posed before her boudoir mirror in a brassiere and slip, would appear to be a bit risqué for either Curtis or the chamber, yet the identifying card in the WSHS carries the notation "this one used." Most of the photographs in the series feature stylishly dressed young women in outdoor settings. In May 1928 Cornelius Banta of the Curtis studio took the photos, the purpose of which is unknown. *Washington State Historical Society, Tacoma, 1943.42.54028.*

The Seattle Regrades

Seattle's extensive regrading of its steep hills allowed Curtis to shoot some spectacular scenes, including this eerie view dated January 25, 1907. The controversial "Denny Hill regrade" is seen from the intersection of Fourth Avenue and Bell Street. The moonscape-like mounds, sometimes called "spite mounds," were the remnants of original lots, the owners of which refused the city's offer to have their property lowered to the new street grade at the same price per cubic yard as the contractor charged the city. The mounds were sloped to minimize earth slides into the newly graded streets. In the center of the photo a "giant" water jet, adapted from hydraulic mining, is reducing some earth to grade. Two large houses, at the left and in the center background, were among those not demolished but lowered from their original elevations to be moved to new locations. *Washington State Historical Society, Tacoma, 1943.42.18733.*

This 1906 photo of the excavation of the remains of the Washington Hotel reveals the depth of the cut at the south face of Denny Hill. The trench at the bottom of the photo is approximately the new street grade. The Washington Hotel later was completely demolished. *Washington State Historical Society, Tacoma, 1943.42.32.*

More typical of regrading was the extensive street work designed to level the hills and hollows on downtown streets, easing the task of horse drawn teaming. Some heavy cuts were necessary, as shown in this view of Third Avenue from Columbia Street on February 3, 1907. *Washington State Historical Society, Tacoma, 1943.42.547.*

Some services were maintained as regrading advanced. This temporary trestle carried the trolley car over the intersection of Third Avenue and Madison Street on January 25, 1907. *Washington State Historical Society, Tacoma, 1943.42.1885.*

The Curtis "Ranch" and Eastern Washington

Curtis captured many of his eastern Washington scenes on or near the small farm he called his "ranch" west of Grandview. Young fruit trees stand, probably on Curtis property, and probably soon after he and Florence purchased it. Stakes designed to hold the trees upright against eastern Washington winds and open irrigation furrows are plainly visible. *Washington State Historical Society, Tacoma, 1943.42.20741.*

This house replaced a small, crude cabin on the property. The photo is dated October 18, 1921. Now within the city limits of Grandview and surrounded by mature trees, the house faces Asahel Curtis Drive. *Washington State Historical Society, Tacoma, 1943.42.42398.*

Unidentified pickers pose for Curtis by an apricot tree. *Washington State Historical Society, Tacoma, 1943.42.44098.*

After picking came packing. *Washington State Historical Society, Tacoma, 1943.42.45712.*

The Chicago, Milwaukee, and St. Paul Railroad commissioned this powerful depiction of a thirty-horse combine in an eastern Washington field. The photo is dated October 9, 1912. *Washington State Historical Society, Tacoma, 1943.42.25660.*

Asahel Curtis and His Family

Florence Curtis on snowshoes and bundled against the weather on a winter 1917 outing to Mount Rainier National Park. *Washington State Historical Society, Tacoma, 1943.42.35770.*

The Curtis children at the ranch about 1923. Clockwise from the left are Betty, Walter, Whitney, and Polly. *Washington State Historical Society, Tacoma, 1943.42.2012.0.47.*

Curtis about the time he left for Alaska, dressed in the mode of a young gentleman. His topcoat sports a plush collar, his suit coat is correctly buttoned, studs fasten his shirt front, and a soft bow tie surrounds his starched high collar. *Courtesy of John Stamets.*

After his return from Alaska ,Curtis was more often photographed wearing eyeglasses. Here he has abandoned the high starched collar for the newly stylish turned down collar and a four-in-hand tie, but has gained a pince-nez. *Washington State Historical Society, Tacoma, 1943.42.20120.16.*

In this snapshot Curtis strides along the reflecting pool in the National Mall with the Lincoln Memorial behind him. He has dispensed with the sartorial elegance and carefully composed facial expressions of his photo portraits in favor of a bulging coat pocket, a camera in his right hand, and a "Between the Acts" cigarillo in his mouth. His friend F. W. "Matt" Mathias may have taken the photograph during June 1936 when both men were in Washington to testify against a proposed park in the Olympic Mountains. *Washington State Historical Society, Tacoma, 1943.42.2013.0.4.*

Curtis in Mount Rainier National Park, atop what appears to be the flat roof of a building. His clothing suggests a time in the late 1910s or early 1920s. *Washington State Historical Society, Tacoma, 1943.42.2013.0.5.*

CHAPTER ELEVEN

Reclaiming the Land, Operating the "Ranch"

IN JANUARY 1907 Curtis bought a nine and one-tenth acre farm from the Washington Irrigation Company on irrigable land in the Yakima Valley, west of the river town of Grandview. The year before, the United States Reclamation Service (the Bureau of Reclamation from 1923) had concluded its anticipated deal with the company, purchasing the canal and other property from the private organization. Federal water would irrigate Curtis's orchards and other crops. Curtis saw no conflict between his belief in limited federal economic involvement and his conviction that, in this case, federal intervention would improve the state's economic outlook, not to mention his own.[1]

As Curtis remembered the transaction, he bought his ranch and became an orchardist after listening "to the alluring talk of E. F. Blaine." Ingratiating, intelligent, and shrewd, developer Elbert F. Blaine sported a flourishing moustache, prominent ears, and deep-set eyes. A New Yorker by birth, Blaine arrived in Seattle in 1885, soon joining the successful law practice of John J. McGilvra. Then he associated with Charles F. Denny of a pioneer Seattle family in the Denny-Blaine Land Company, the developers of the Denny-Blaine Lake Park Addition and other properties. No armchair developer, Blaine, probably with Denny, paddled the twenty-five mile Sunnyside irrigation canal in a canoe before deciding to invest in the Washington Irrigation Company and expand its canal. Blaine had no intention of keeping the company in the irrigation business. In his view the irrigation company was a device for raising land values on monetarily worthless sagebrush desert. Irrigation itself made little money, but selling intensively cropped farm land on rising values made a lot.[2]

A brief look at the development of irrigation in the arid West reveals why Blaine and his partners were uninterested in the indefinite operation of their irrigation company. To begin with water rights, the state of Washington applied the common "doctrine of prior appropriation" to water use. Thus the first water user along a stream was "first in time" and therefore "first in right." The doctrine did not apply to the abundantly watered coastal strip of western Washington, but was overwhelmingly important in parched areas of eastern Washington, where summer temperatures could top one hundred degrees, and where rainfall ran about seven or eight inches annually, most of it falling in the winter, when it was of little use for crops. Any user's right to water was limited to a "beneficial" use, but irrigation of trees or other crops was a "beneficial" use, insulating "senior" or "first in right" irrigators from most formal challenges. So far so good for the irrigator.[3]

Alas for the irrigator, the practice of irrigation was less straightforward. The irrigator almost had to have monopoly on water use for a prescribed territory, irrespective of whether the monopoly was private, like Blaine's, was a government entity, or was owned cooperatively by the landowners who benefitted from the flow of water. Only the monopolist could decide vital questions about the "duty of water," or the volume necessary to produce crops on any parcel of land. Related issues included how the interests of each landowner would be satisfied within the context of the whole acreage to be watered, and how the entire volume of water would be determined. A monopoly was rarely "first in right," and therefore remained under legal or quasi-legal pressure from prior claimants. Even if "first in right," the monopoly was subject to informal demands for more water from other users and from its customers. Any monopoly bred resentment, which could be freely expressed in the farming communities of the United States, and beyond that, the monopolist had many responsibilities. The first was for the "headworks" designed to direct water from a natural stream into a main canal or "lateral." Next came the canal itself, paralleling the stream on a gradient permitting excess water or waste water to return to the stream at some point below the irrigated lands. In a simple situation, the individual farmer turned a valve diverting and measuring water from the monopolist's canal or lateral to his own lateral, from where it ran into furrows roughly right-angled to the farmer's lateral. The irrigated crops grew on either side of the furrows. If the monopolist controlled a lake and a dam at "his" lake, he could operate a "high line" conduit to irrigate slopes too high above a stream for water to reach, except by pumping.

Irrigation operations were simple in theory but complex in practice. All irrigated acreage had to be prepared to achieve the correct balance between land level enough for efficient farming, yet sloped sufficiently for the water to run downgrade along each farmer's lateral and furrows while percolating into the soil at an approximately equal volume for each tree or plant. High winds that blew away topsoil, seeds, or young plants called for replanting and extra watering until the roots could reach down into the furrows. The farmer had to maintain his furrows and lateral scrupulously, repairing any leaks or breaks in their banks at once. Laterals and furrows could foul with debris, upsetting their carefully calculated sloping, forcing water to back up and overflow. Only quick work with spades and shovels would keep water moving downgrade. Farmers feared too little water, but too much could result in significant crop damage. Serious overwatering raised the water table. As water rose, it at times brought previously undissolved salts, collectively called alkali, to or near the surface. Evaporation could dry the water but leave the salts in deadly concentrations, burning tender roots and killing plants.[4] Farmers had other problems. They were subject to the vagaries of competition, disorganized marketing, cycles of overproduction and plunging prices, freight rates (always too high), and crop losses.

Back at the monopolist's headquarters, the uninterrupted performance of headworks and the main canal or lateral (or laterals) was absolutely essential. The monopolist had to decide how much to charge each farmer for a "water right," and how much to assess him for maintaining the monopolist's equipment, and for covering other expenses such as salaries and depreciation. A farmer could refuse to pay. The monopolist could shut off his water but an inexpensive valve or gate from the monopolist's lateral to the farmer's lateral could be tampered with. The monopolist's alternative was to hire men to watch the gate twenty-four hours a day to gather evidence of tampering that would stand up in court. The task was nearly impossible if farmers sympathetic to the non-paying farmer tampered with other gates up and down the lateral in the wee hours of the morning.[5]

Which returns us to the Yakima River, the Washington Irrigation Company, the Reclamation Service, and Asahel Curtis. The Yakima and its tributaries rise in the eastern slopes of the Cascade Mountains or their foothills. The river moves along more than two hundred miles southeasterly to debouch into the Columbia River at Kennewick. Melting snows from the mountains and foothills feed the river. Some of the melt now is held in Bureau of Reclamation reservoir lakes to compensate for the normally low water and scant rainfall during the valley's hot summers.

When Curtis bought his ranch, the country had put the Depression of 1893 with its "terrible nineties" behind it. Yet relatively good times did not bring efficient farming to the Yakima Valley. There were no reservoir lakes. Water flowed directly from canals and laterals to the furrows during the summer when the Yakima was at its lowest. The company and its customers discovered in 1901 just how inefficient these arrangements could be, when the threat of a lawsuit from the holder of a superior water right forced a curtailment of the company's water flow. Low water struck again in 1903, 1904, and 1905, unfortunately during late August, when crops were most dependent on irrigation.[6] The shortages were not the company's fault but farmers blamed the company anyhow. During the 1905 crisis, when the company's manager talked to some users about restricting their water intake to allow other farmers adequate amounts, they refused. After he installed measuring devices to limit consumption, in the dead of night men broke the small dams leading water to the devices. The complex irrigation system of several farms on each of several sub-laterals prevented the company from pinpointing the malefactors.[7]

Worse for the company, while crops on some irrigated land shriveled and died, other land suffered inundation. If that makes little sense, consider two facts. First, irrigation raised the water table on lands once given over to "lizards, jackrabbits, coyotes, and rattlesnakes," from forty feet below the site of Sunnyside town in 1883, to ten feet below in 1903. Seepage from higher elevations to lower exacerbated the problem in low lying farms. Some were lost to flooding and alkali. In 1904 the water table rose above the low places. Second, the Washington Irrigation Company never provided adequate drainage to funnel excess water from cultivated lands back to lower stretches of the river. The drainage problem was worse than it had to be, for most water users consumed more than they needed, in some cases to validate their extravagant water right claims. Surplus surface water and rising ground water ran through river towns, turning unpaved streets into quagmires. In Sunnyside, viscous streets mired buggies and wagons. Water rose through depressions in the streets, but loads of rock dumped to fill the holes only forced the water up to make new sinks elsewhere. Basements flooded. Any excavation dug for a basement quickly became a miniature lake. For a time it was next to impossible to move anything in or out of the town. The first Northern Pacific train to Sunnyside foundered on its submerged tracks, flipping on its side.[8]

The Washington Irrigation Company was scarcely the overmastering monopolist abetting the landed fiefdoms depicted in *Rivers of Empire,* historian Donald Worster's study of irrigated farming in the southwest. In fact the weak company could not do anything about its situation on the ground, so it tried to meliorate its position and improve its future prospects through politics. E. F. Blaine ushered a bill through the Washington State House of Representatives allowing the company to build a reservoir at Lake Cle Elum, northwest of Sunnyside, but the maneuvers of a hostile senator killed it in the upper house. Next the company tried to revive the Carey Act of 1894, never an effective piece of national legislation, and practically moribund after the passage of the Reclamation Act of 1902, of which more later. The Carey Act, not repealed by the Reclamation Act, remained on the books. Under its terms a western state could receive as much as one million acres of public land if it promised to reclaim the land or arrange for private parties to do so. The company's request under the Carey Act opened the possibility of state action to improve its supply and drainage circumstances through the reclamation of 56,000 additional acres. Presumably the company could expand its operations enough to afford abundant water and adequate drainage canals. Anti-monopoly opponents contested the application on two divergent grounds. They objected, first, because they doubted that adding Carey lands would make any more water available, and second, because placing Carey lands under the company would preclude any other private development. The opponents delayed approval long enough for the company to embrace another solution, selling its irrigation works.[9]

It was evident to the company and to its one possible purchaser, the Reclamation Service, that the company was at the end of its irrigation-and-water-disposal tether. Yet the company's inadequate intake and outflow works by no means suggested stagnation in Yakima Valley development. The Sunnyside Canal and its branches grew from 400 to 500 miles in 1901 to almost 700 miles in 1904, when about 36,000 acres were irrigated, or "under the ditch," as the phrase went. In 1905 some

5,000 people lived on properties watered from the Sunnyside, around 20 percent of the valley total. The population of North Yakima (later Yakima), grew from about 3,000 in the depression year of 1893 to almost 7,000 in 1905.[10] More growth, however, was problematic unless substantial investment capital could be found to expand the company's works.

Enter the Reclamation Act of 1902 and the Reclamation Service. The Reclamation Act also is known as the Newlands Act for its sponsor in the House of Representatives, Congressman (later senator) Francis G. Newlands of Nevada. The act provided for federal irrigation works on irrigable land in the western states. The Reclamation Service, created to administer the act, would control settlement on the public lands set aside for reclaiming, to ensure that actual settlers took up the land, and that no speculation on future land values could occur. Privately owned land was eligible for reclamation if its owners agreed to accept federal limits on farm size and speculative profits. Settlers were required to form associations to distribute water as far as each farm, where the owner would build and maintain his own laterals and furrows. The settlers were required to repay the costs of the irrigation works within ten years, after which all of the works except reservoirs would revert to their associations. Public land sales would create a revolving fund to establish new projects.[11]

No such far-reaching act springs from a void. The Reclamation Act arose from a complex mix of analysis, belief, and emotion, but the strands may be disentangled. First came the indisputably grim reality of reclamation. There was no future for reclamation because the private capital to expand irrigation into rich but arid soils simply was not there, despite the generally good economic times of the early twentieth century. Only the federal government had the resources to promote further reclamation. Yet the lack of private investment capital merely begged the question, forcing reclamation's proponents to answer the retorts "so what" or "why bother?" In other words, what justified reclamation enough to spend federal money on it and, as some opponents suggested, eventually put the taxpayer on the hook for the advantage of a relatively few farm families?

One answer was that the arid West deserved settlement as much as the humid east. As famers played a significant role in eastern growth and development, so should they be encouraged to settle the underpopulated West. Moreover, the expansion of farming was not keeping pace with population growth. Therefore more farms were needed to stave off agricultural imports, which lined the pockets of foreigners and transportation companies while they diverted potential income from American citizens.

A belief in the superiority of farming as a way of life formed another strand of pro-reclamation advocacy. Here belief and emotion mingled, for there was no absolute proof of the assertion that farming life was finer or better than urban life. For the advocates of rural living the proposition needed no statistical buttressing. The farmer was more self-reliant, more independent than the city dweller, more content to live close to the soil, rather than close to the pavement. The soil, after all, was the ultimate source of wealth. Farmers were interested in their communities through their property ownership, in contrast to the dweller in a tenement house. Property ownership invited critical thinking about social and economic policies affecting property, and so, by extension, about strengthening the mutually reinforcing supports for property located in the broader culture. Rural property ownership was conservative in the best sense of the word, because it buttressed human values rather than materialistic striving. It produced the psychic benefits of contentment, serenity, mutual respect, and happiness.

"The Yakima Valley is essentially a region of prosperous, happy farm homes," declared a 1936 study by the Washington State Planning Council (WSPC) designed to parry demands for curtailed reclamation. The italics are in the original. *"To the average Yakima Valley farmer, farming is a mode of living, rather than a business,"* the study continued.[12] Those assumptions formed the core of the "back-to-the-land" movement, significant in Curtis's opposition to the massive Olympic National Park. The movement itself contained three separate strands: the assumed unhealthy growth of big cities and lesser but still large urban agglomerations; the potential social dangers arising from dwellers in cities who retained their traditional cultures; and an emphasis on the counterbalancing effect of rural resurgence. As the twentieth century advanced, the back-to-the-land emphasis shifted from promoting individualism and self-reliance through land settlement, to fostering community cooperation in planned farming or mixed farming-and-industrial developments. Belief in the superior values of rural life persisted, reaching the cooperative apotheosis during the federal New Deal programs of the 1930s, such as the Resettlement Administration.

The sheer growth of huge cities was a stunning phenomenon. By 1870 New York City and Brooklyn, later a part of the city, were well over the million mark. By 1900 the consolidated New York was pushing 3,500,000. New York was gargantuan, of course, but a long way from the Yakima Valley. Closer cities were scarcely tiny. In 1900 Chicago boasted almost 1,700,000; San Francisco close to 350,000, and Denver almost 134,000. These were large numbers for the time.[13] By 1910, about four years after Curtis bought his Yakima Valley property, fifty cities would pass the 100,000 mark, including Seattle at over 237,000.[14]

Other trends were at least as disturbing for the advocates of rural life. In 1910 the country's urban population reached 45.7 percent. Despite the U.S. Census definition of an urban place as a mere 2,500 or more, the direction of growth was unmistakable.[15] Equally unfortunate from the back-to-the-land perspective was the large percentage of foreign born and of native born people of foreign born or mixed native and foreign born parentage. In burgeoning New York the percentage was 78.6, but New York was, well, New York. In Chicago the percentage was almost as high, 77.6, and in Cleveland—what city could be more Midwestern than Cleveland—it was 74.8.[16]

Many of these urban residents were products of the so-called "new immigration" of people from southern and eastern Europe whose customs and languages were not English, or adaptable western European cultures or tongues. The reality of internal migration, of Americans leaving farms for, if not big cities, at least for regional cities, did nothing to assuage the fears of rural advocates that fundamental American values were under siege. For them, the decline of America could be traced to the turbulent, crime ridden, and corrupt cities.[17]

The concern over rural recession and urban dominance was real enough, fed not alone by anxieties over urban evil, but also over the disappearance of a demarcation between settled and unsettled land, the so-called passing of the frontier. In this view there was no coincidence between the formal closing of the frontier in 1890 and the devastating economic depression that soon followed. Unless new agricultural frontiers were opened, the catastrophic depression could recur. Historian Donald J. Pisani identifies these convictions of the federal reclamation advocates, who "sold the policy to the public as a program to salvage the 'wasted' barren lives of homeless city dwellers and to arrest a perceived decay of republican ideals and civic virtue."[18] The public bought what they were selling. Western railroads and the hyperactive President Theodore Roosevelt backed the Reclamation Act, signing it into law on June 17, 1902.

At first the Reclamation Service was uninterested in the Yakima Valley because there were plenty of other potential irrigation sites on public land. It reversed course because the Washington Irrigation Company wanted to sell, because the Washington legislature passed an enabling act, and because the Yakima Valley works could show results right away. The sale closed March 28, 1906. Settling the messy business of tangled water rights followed. The Reclamation Service built a new diversion dam for the 1907 irrigation season. It finished its first of several reservoir dams in 1910. The drainage issue was not met until after 1913, when a state law provided for the assessment of farms and other improvements in drainage districts to build and maintain adequate drains.[19]

It's doubtful that Curtis bought his land, just before the federal improvements came on line, only because of Elbert F. Blaine's "alluring talk." Tying his purchase to Blaine, by then a prominent valley resident, and to an early emotional commitment to land holding and orchard development would go down well with fellow orchardists. While Blaine's sales pitch and Curtis's sentiment may have played roles in the decision to buy, careful calculations undergirded his venture.

First, Curtis believed in the back-to-the-land movement. He emphasized, years after his purchase, the advantages of rural life rather than the deleterious effects of cities, for he was an urban man. He wanted Seattle to grow. He wanted all of Washington to grow. He wanted the whole Pacific Northwest to grow. One healthy way to encourage that growth was to keep at least some people on the land, and arrange for others with agricultural inclinations to take up rural living, though not all of them could, or else Seattle and other cities would not expand. Curtis himself was no stranger to farming and gardening. He was involved with farming and with plants from his pre-teen years, first in Minnesota, then at Port Orchard during his teens. "He loved his garden," his daughter Betty recalled. Working in his garden "was his greatest relaxation away from the office."[20] He probably had gardens at each residence, though Betty most likely referred to a fully developed garden he created after the family settled in the Belmont Avenue house. But wherever it was, his garden was another affirmation of his zest for nurturing plants.

Second, Curtis was prospecting for new photography business. One way to get it was to demonstrate a commitment to reclamation and irrigation in eastern Washington, and there was no better way to assert an interest than by buying irrigable land. As historian G. Thomas Edwards demonstrated, at least as early as 1905 Curtis was taking photographs of valley scenes for the promotional booklets prepared by commercial groups and sent to prospective settlers and others.[21] It is possible that he wrote all or part of the copy for some booklets because he had a talent for converting facts and statistics into fluid prose.

Third, Curtis was a robust thirty-two, young enough to manage a small farm, even an intensively developed one, with family or hired help at planting and harvest times, and specialized assistance at others, as with pesticide spraying. The farm could also be a temporary escape from Seattle, as well as a place from which to seek out business opportunities or fulfill business obligations. Curtis joined what C. Brewster Coulter identified as the fifth of five waves of settler migration onto the valley farms through 1910. This "fifth wave" came during the boom that "followed the Reclamation Service into the valley."[22] The members of the fifth wave called themselves "book farmers," people who had been "clerks and mechanics" but "were learning the latest technical and scientific advances out of books and learning them rapidly."[23]

Curtis was no "book farmer." He was a seasoned grower, tender, and harvester of crops, even if not well acquainted with irrigation. Given his determination, dexterity, and physical condition, he would have learned irrigation quickly. Dated photographs of his ranch show it as at least partly developed and "under the ditch" when he bought it.[24] It was a testimonial to Curtis's energy and stamina that he was climbing mountains, organizing expeditions, running a busy photo studio, promoting Mount Rainier National Park, and raising a family while he was developing his property. Soon he would be trying to carve an Olympic national park from the Mount Olympus National Monument, organizing the Rainier National Park Advisory Board, serving on committees of the Seattle Chamber of Commerce, and working to advance the struggling good roads movement.

It isn't known how Asahel and Florence juggled the demands of the valley acreage with family life in Seattle, not to mention the studio and all of Curtis's other activities. There is some evidence that they were not always successful, as when Florence and Betty spent some time in Portland and Curtis made his second, extensive trip to and through Alaska. Betty remembered that her first summer sojourn at the ranch happened in 1915, when she was four years old. From 1915 until 1922 or 1923 she spent every summer there with Florence and her brothers. Her recollection omits any mention of Polly, born in 1920. "We lived in a small three room cabin—and I do mean cabin—Mother, my two brothers and I." Asahel was there only on weekends because dust and possibly pollen allergies and high temperatures made longer stays impossible for him. His reactions moderated during fall harvest time, so he could remain for longer periods. The annual household move from Seattle to Grandview made no special provision for Florence's pregnancy or Polly's birth, at least not in Betty's recollection. Evidently Florence had some help coping with her growing family, but her summertime lot could not have been an easy one. With school out, she had to organize a railroad trip of about two hundred miles with two, three, or four children and their paraphernalia. She had to supervise her brood while in the Yakima Valley, then return to Seattle in time for school in the fall. After the children were in school, "Mom returned to take care of the apple harvest."[25] There had to be some help with the children in Seattle during harvest time, perhaps from Asahel's sister Eva.

A new house on the property, first photographed in 1921, may have eased Florence's parental burdens.[26] The house was a story and a half frame bungalow with a shady front porch, built to accommodate the larger family. Certainly Whit and Walt, growing into their teen years, required more space. Florence's obligation to organize family trips back and forth from Seattle to Grandview continued despite any improvement in living conditions. Her accumulated experience may have mitigated the annual logistical imposition, but it may also have meant putting Betty in the Grandview grade school for a year. And not only Betty. "I think my brothers stayed over a couple of times,"[27] she recalled. Neighbors or hired help would have filled the gaps in supervision between parental stays.

The adult difficulties with maintaining two households were of no concern to young Betty. "We all worked some around the place of course but enjoyed the farm life too." There were horses and the usual childhood scrapes. A second-hand Model T Ford graced the ranch. Betty learned to drive at age eleven, and perhaps her brothers mastered the

Model T at younger ages. Early driving was common among farm children at the time, and all the more essential because Asahel did not drive. Betty couldn't "see how I could reach the pedals but we did."[28] More than reaching the high-low gear, brake, and reverse pedals was involved, because the children had to absorb the intricacies of the spark and accelerator levers, and the parking brake-neutral lever, as well as the tasks of driving. Unless the Ford was fitted or retrofitted with a battery and electric starter, unlikely accessories in an older model, the motor was hand cranked. Hand cranking was potentially dangerous because a "kickback" could break a wrist. If the task was beyond Betty or her brothers were not around, Florence started the car.

There were other diversions because Whit and Walt played music for dances. Betty did not mention their instruments but did assert that "we all went—down at a place called Mabton Park on the Yakima River." The hours there were filled with "bugs, watermelon, fried chicken and fun." Rural life for the Curtis children was exactly what the back-to-the-land advocates claimed for it. "Actually," Betty remembered, "Ive [sic] always felt it was probably the best place in the world for us all to grow up. We were a pretty frisky bunch and this was a wonderful outlet."[29]

Children and their summertime fun aside, the farm's principal function was crop raising. Curtis's ranch, intensively farmed, was sufficient for a profitable operation, at least in some years. The spread had other purposes than money making, so its small size, a little more than nine acres versus an average of thirty-five in Yakima Valley farms, was not critical. Betty recalled how harvests at the farm and the photography business balanced out. When "the office" hit "a low point we had a good apple year and vice versa."[30] Winesaps and yellow newtons were the mainstays of the apple crop. Both were good eating and cooking apples, and old, proven varieties. Other types supplanted them in later years, but their lineage and dependability would have appealed to Curtis.[31] Apricots, grapes, and potatoes planted between the tree rows were lesser crops. There was a kitchen garden. Valley farmers cultivated alfalfa, a forage crop for livestock, but Curtis probably did not grow it because of his limited acreage. Betty remembered how the kitchen garden shared space with plots for "a bunch of 'goofy things' Dad tried—grafts etc."[32]

In 1921 and probably in other years Curtis donated a truckload of apples to the Ryther Home in Seattle, a shelter for orphaned and abandoned children operated by locally famous "Mother" Olive Ryther. "Ollie" Ryther enjoyed an enthusiastic following, and it is not surprising that Curtis counted himself in the group. Two photographs record his gift. One dated October 18, 1921, shows a truck loaded with apples, while the second, from October 29, displays a pile of apples with some residents of the home.[33]

For all the side activities, it bears repeating that the ranch did not exist primarily for charitable reasons, or as a children's playground, or as a base for other business. It was a working farm expected to make money. For that reason, rampant individualism had no place in Curtis's vision of profit. No money could be made in an intensely competitive environment with each grower attempting to undersell the others, so savvy cooperative marketing was essential to each orchardist's economic health. The concept was nothing new but a permanent marketing organization did not emerge for many years. Curtis's commitment to the movement is evident in two photographs. The first depicts a 1910 display of boxes marked "Yakima Valley Fruit Growers Association," an early effort to achieve the cooperative ideal. The later photo (see page 48) shows four-year-old Betty, dressed in a suitably dirty frock, supposedly assembling a wooden apple box. In her left hand she holds a nail, about to pound it into a top slat and the upper edge of one side of the box. Her right hand holds a hatchet, its blade up, its hammer head ready to drive the nail home.[34] The photo was posed. Curtis expected the children to work around the farm but a four-year-old hammering with a hatchet in a box assembly area surely was beyond his requirements. Yet the picture made two points; farming was a family enterprise, and standardized shipping containers were a necessary part of successful apple marketing.

Curtis abandoned active management of the ranch in the 1920s. "After about 1925 or so," Betty recalled, "the place was rented out to Alfred Jensen."[35] There were reasons why the place became too heavy a burden. Whit and Walt were grown. Betty, in her teens, was probably less willing to leave her peers for a summer away from Seattle. In the prosperous later 1920s the photography business demanded Curtis's time and attention. His commitment to Yakima Valley photography and his growing role in producers' trade associations took much of his time. In the 1920s his fights with Governor Roland Hartley over highway funding intensified, as did his struggles for increased Rainier park appropriations and for constructing the West Side Road.

Florence's mental instability likely was another. "Mrs. Curtis is seriously ill and I will need all my personal finances to pull things through," he wrote in May 1928.[36] He was responding to a request to renew his lapsed membership in the Washington State Chamber of Commerce and remain on its board of directors. While it is possible that Florence's problem was physical, "seriously ill" suggests a disease, not a condition such as cataracts. There is no mention of surgery. To be sure, people in Curtis's era were more reticent about any deviation from physical or mental norms than they were in later times. The reluctance to be forthcoming about such matters was especially evident in the Curtis family, so the letter is reserved. If Florence's difficulties were mental, they may have involved her institutionalization, an expensive program. Whatever the exact circumstances, Curtis in his fifties was too involved with his business, Rainier, good roads, and apple marketing to spend his declining energy either in farm labor or on less pressing issues involving the state chamber of commerce.

The ranch, nevertheless, appears to have done well enough as a leased property until 1931. Then the Grandview Cold Storage Company paid the taxes on the property and continued to do so for many years, even after Curtis no longer owned it. In January 1935 the property was sold at public auction to pay a delinquent assessment levied by the Sunnyside Valley Irrigation District. In February the buyer transferred the sale document to the Perham Fruit Company, a firm owned by the owner of the Grandview Cold Storage Company. In March Curtis and Florence mortgaged the property to the Grandview Cold Storage Company, declaring on the mortgage contract that the title to their property was "valid and unencumbered." In January 1936 the time to redeem the property expired, and the Perham Fruit Company assumed title.

The bare bones of these transactions leave questions unanswered, scarcely a unique situation in Curtis's life. Betty knew something of the matter because she was helping with the photography business when the ranch was in its final struggles. She recalled that the property was foreclosed because it failed to generate the income necessary to repay federal loans, surely a false memory. She did recall her father's receiving "a check for a couple of hundred dollars…and the ranch wasn't his anymore." Betty remembered the incident "so well because he looked at the check and asked what the Hell he should do with it—maybe go on a good drunk, which sure surprized [sic] me because I'm sure he never had more than three drinks at a time before then."[37] Curtis did not squander the money on drink despite his loss, yet "it was sort of a big let down to have most of a life time of work and struggle gone like that."[38] Betty's strong recollection of the check and her father's anguished disappointment make it likely that Curtis emerged with some money from the strange series of transactions.

Meanwhile, Curtis worked to solve apple marketing problems. He limited his marketing efforts principally to establishing an effective organization embracing all the people involved in marketing. Growers, middlemen, shippers, advertising agencies, and retailers had to work in harmony to provide a fair income distribution for all. They needed to develop markets. They had to find a way to maintain prices while additional irrigation projects came on line in the Yakima Valley and elsewhere. The new projects were designed to lure families back to the land but also, necessarily, had to provide their settlers with fresh opportunities for expanding commercial sales. It was a tall order, one not partially filled in Washington until 1937 with the legislative creation of a state agency with enforcement powers, the Washington State Apple Advisory Commission, or as it was usually known, the Washington State Apple Commission. Meanwhile, Curtis and like-minded growers strove to find a way to stimulate sales while assuring fair returns to all participants in the marketing process, especially themselves. Curtis wrote few comments about early attempts to attract consumers locally, regionally, or nationally, but there isn't much doubt of his approval. At the same time he surely understood the limits of uniform grading, labeling, and packing, small voluntary cooperatives, fruit expositions, apple shows, and product exhibition cars on national railroads. Each of them was helpful, to be sure, but they were not coordinated with other programs and were subject to changes and cancellations.[39]

Curtis took the lead in founding a new, comprehensive organization. In January 1925 he chaired a conference called to promote apple sales, but the conference decided that standards for grading and packing the fruit should come before sales promotion. Progress in defining uniform standards led Curtis, in April 1926, to call together a group to organize what became Pacific Northwest Apples Inc. He acted as chair of the meeting while Thomas B. Hill served as secretary. Curtis and Hill served in analogous positions on the State Development Committee of the Seattle Chamber of Commerce. They received their impetus from the extension service of Washington State College, which suggested a proven method of financing marketing and advertising by placing small charges on shipped boxes and loaded freight cars. Curtis appointed committees to gather ideas and information from various apple districts. That work completed, the group decided at their next meeting on June 26 to begin with a small advertising campaign. Modest fees on boxes and carloads, plus the sale of the corporation's stock at one dollar per share would finance the publicity. Curtis and Hall, having set things moving, dropped out of the organization's hierarchy. A Seattle advertising agency associated with the corporation published a brief book in the same year rehearsing the problems of the apple industry—lack of cooperation, high freight rates, cultivation issues, wildly fluctuating prices, and so on. There was hope, nevertheless. Pacific Northwest states shipped more apples to the thirteen largest cities than strong competitors in Michigan and New York. Pacific Northwest Boxed Apples Inc. should therefore concentrate on consumers in the United States rather than trying to pry open "markets in far-away lands."[40]

Unfortunately, when the corporation attempted to make detailed plans, dissention surfaced. In the spring of 1927, Curtis asked for a tentative plan to be put in writing, and was rebuffed. "It would appear that the plan is a sacred thing that the ordinary human intelligence cannot comprehend and something too holy to be discussed," he wrote in frustration. Instead of producing something tangible, he complained, people criticized him for demanding details. A "feeling on the part of the growers that we were trying to impose an organization from above" also hampered progress. By October 1927 it was clear to Curtis that the corporation was going nowhere despite having about $12,500 on hand, despite the earnest work of individuals, and despite his paying part of Hill's salary and expenses so that Hill could devote time to organizing. While the corporate leadership dithered, a successful campaign to promote Jonathan apples led to the formation of the Washington Boxed Apple Bureau. Pacific Northwest Boxed Apples Inc. went out of business. The Washington Boxed Apple Bureau succeeded until the Great Depression, when many of its members could no longer pay their assessments. Its legacy of effective organization lived on to inspire the state agency, the Washington State Apple Advisory Commission.[41]

The necessity for any successful growing and marketing scheme was, of course, irrigation. Curtis was not a founding member of the Washington Irrigation Institute, organized in 1913 to promote settlement on irrigated land, but he joined the organization in 1920. The same year he spoke to the annual meeting, showing "most wonderful" slides in an illustrated lecture. The institute was small, not more than a few hundred members at most, but it was active. C. Brewster Coulter found that "the Washington Irrigation Institute and the Washington Horticultural Association [which Curtis did not join] were the most effective vehicles for educational self-help among the farmers" of the Yakima Valley.[42] The Institute "unanimously elected" Curtis president at its 1928 meeting, its "most successful convention of the past several years." Though he preferred to see the presidency in the hands of someone from across the Cascades, he agreed to serve.[43]

The next year he delivered his presidential address at Yakima. Rural primacy over cities formed his theme. He told the nearly one hundred delegates that "reclamation has built many rich, beautiful cities as Yakima and has made 'give us this day our daily bread' little more than a figure of speech." Prosperity and security were dependent on farm production because "the products of the farm have never cost so small a proportion of the family income. We forget that the grim shadow of famine has darkened man's pathway through the ages," a shadow lifted partly through reclamation. He responded to reclamation's critics who told farmers, when "farm profits dwindled… that we were fools for ever having turned a furrow." In reality, reclamation had to continue with increased attention to federal-state cooperation, research, marketing, standardized methods, resource development, and planning for the future. "The agricultural resources supporting our cities are as important as the raw materials used in manufacturing."[44]

The convention provided some hearty laughs, partly at Curtis's expense. Ralph Williamson, a friend and fellow Republican, told a dinner meeting that Curtis's bald head reminded him of the new reflector at the Mount Wilson observatory. Curtis interrupted Williamson's talk to compare Williamson's head to his grandmother's "silver coffee urn," whereupon "the fun started… Before the evening was over a bald-headed club was organized with Williamson, Curtis, Worrall Wilson and R. K. Tiffany as charter members." The 1929 convention may have provided the last amusement Curtis enjoyed from reclamation.[45]

Curtis's final recorded involvement in the institute came at the 1931 annual meeting in Wenatchee, where he chaired and spoke at a meeting of the resolutions committee. He confessed that when federal reclamation began in 1902 there was relatively little knowledge of irrigation's pitfalls. "During these years we have had to part with some of our illusions," he admitted. Among other suggestions, he called for protecting watershed areas and flood control. Irrigation districts were charged for reservoirs but Yakima Valley towns benefitted from flood control as much as did farms, yet paid no assessments for water storage. They should be assessed. Above all, the institute should recognize that reclamation's focus should be on irrigation and drainage. "The land is the least valuable element in the equation; the water and the settler are the most valuable." Land should remain cheap through a system of classification and appraisal designed to curb speculation by effectively prohibiting sales above the appraised price.[46] His talk was apparently his final appearance at institute meetings, for he is not listed as a member in the faltering organization after 1931.[47]

Curtis maintained an interest in irrigation development in the Yakima Valley despite the fading Washington Irrigation Institute, the loss of his ranch, and calls to curb reclamation because of agricultural overproduction during the Great Depression. There was an economic motive behind his continued interest. He took photographs for the, by then, Bureau of Reclamation. His commercial customers included, among several others, the Washington Boxed Apple Bureau, the Kittitas irrigation unit northwest of the Sunnyside unit, and the Roza unit, a "highline" or pumped water development paralleling the Sunnyside unit.[48] Lining his pockets was not his only motive, for he believed in irrigation farming as an answer to the Great Depression. Irrigated land produced few crops such as corn, cotton, and wheat that were in serious surplus, therefore irrigation could proceed. Besides, the Great Depression reversed the country-to-city migration so apparent since the later nineteenth century. In 1930, 17,000 people moved from the city to the country. The next year 214,000 followed. By 1932 the tide became a flood, with an astonishing 533,000

Americans abandoning the city. They had to lead productive lives somewhere. Why not on irrigated farms? Curtis may not have absorbed all of the arguments in favor of irrigation, or the statistics of the unprecedented return to the land, but he surely knew of the general pro-irrigation assertions and of the broad population movements.[49]

Curtis's relations with Elwood Mead, the dynamic commissioner of the Bureau of Reclamation from 1924 until his death in 1936, were generally cordial. Mead's insistence on a state guarantee "of at least $400,000" to be used for the Kittitas unit was a fly in the ointment of friendship. In defense of Mead, his personal history of advising land settlement projects left a string of failures in his wake. Besides, not every Reclamation Service-Bureau of Reclamation project was going well. Curtis was correct to criticize Mead for "his lack of understanding of the situation in the Kittitas," because, when opened in 1932, "the Kittitas Division" was "settled up more rapidly than any of the earlier units of the Yakima project." The eagerness of "nearly all" of the Kittitas settlers was traceable to their "previous experience in irrigation farming."[50]

It was all very well to criticize Mead after the event, but the WSPC's hopeful analysis of the irrigation situation from 1929 through 1933 scarcely was reassuring. The planning council found that the number of "establishments"—businesses in the six major irrigated counties—numbered 266 in 1929, and 176 in 1933, or a little less than a 33 percent decline. The number of wage earners in the counties, 4,589 in 1929, had fallen to 3,693, a shrinkage of a bit more than 20 percent. Yet those numbers and percentages were less comforting when plummeting wages were considered. In 1929 the wage earners were paid just over six million dollars, in contrast to 1933's wages, almost three million dollars less. Thus while the number of businesses and wage earners had held up reasonably well, the businesses and the farmers were paying in 1933 less than 51 percent of the wages of 1929. The worth of the products of the six counties told a similarly sad tale. The value in 1933 was a shade less than 52 percent of the 1929 value of $28,357,747.[51]

What was worse, apple production showed no reduction in response to lower returns. During the seasonal year 1928-1929, a total of 3,279,711 apple boxes shipped from Seattle, while the 1933-1934 year saw 4,311,623 boxes shipped. Annual fluctuations were related, apparently, to the quality of the fruit and the quantity of the harvest, so reduced prices did not result in lower production. This reality made the situation all the more lamentable for the six counties, despite the council's insistence that they were economically better off than the rest of the state. The council marshaled evidence for its conclusion but only served to highlight the dismal conditions in the rest of Washington.[52]

For Curtis, however, irrigation's ills could be cured by more irrigation. His only proviso was a more thoughtfully constructed reclamation, including land use controls, careful selection of settlers, curbs on speculation, provisions for rational marketing and state or federal financing for existing, financially strapped irrigation districts, and, by extension, suffering farmers like himself. Despite the difficulties, he favored expanding reclamation into the Columbia basin east of the Yakima Project.[53] The fight over the Columbia basin involved not whether it would be irrigated, but how. As historian Paul C. Pitzer has shown, the battle pitted the "pumpers" who wanted a dam on the Columbia at the head of the Grand Coulee against the "gravity men" who plumped for diverting water from Idaho's Pend Oreille River downhill to the Columbia basin. The dam on the Columbia would take water from the Columbia and pour it into the Grand Coulee, from where it would be pumped up to a system of canals in the Columbia basin. Curtis was an honorary trustee of the Columbia Basin Irrigation League, and therefore a "gravity man" closely allied with the Washington Water Power Company, the private concern angling to build the gravity system.[54]

Curtis differed from his fellow gravity men in his comprehensive approach to the problem. He argued "that we cannot develop the area until we have come to a thorough agreement with the other states," meaning Idaho and Montana, "after a complete investigation of the uses of the waters of the Columbia," for "reclamation, power and flood control." In this stance he followed Herbert Hoover, whom he greatly admired. On August 19, 1926, Hoover, then Secretary of Commerce and in favor of the gravity idea, reviewed the site at Albeni Falls on the Pend Oreille, where the gravity men planned a low diversion dam. Engineer Hoover approved the location. At a speech in Spokane that evening he declared the gravity system "inevitable" but warned that the inevitable would take time. The next day the Spokane *Spokesman-Review* ran a photograph of Curtis, Hoover, and another man deep in conversation. Curtis cherished the photograph. On the 26th, Hoover spoke in Seattle before a crowd of 10,000 at the University of Washington stadium. He favored, he said, combining the reclamation of up to 1,750,000 acres in the Columbia basin with flood control and power development. Flood control and power were not part of the private gravity plan, so it was just as well for the gravity men that Hoover embraced a wider system at Seattle, not Spokane. His vision transcended immediate issues when he called for "a national program of water development" not to be "delayed until we are overwhelmed with population in this country." Water that was not producing something was a lost resource, he declared. His speech "provoked a demonstration" from the crowd.[55]

Despite the enthusiasm for the gravity plan it gradually lost ground to the pumpers and the concept of a vast diversion dam on the upper Columbia. When President Franklin D. Roosevelt took office in 1933 the question was settled in favor of a dam and power plant. By the summer of that year Curtis feared the possible overexpansion of irrigation but hoped "to find some way of utilizing the power." A 1935 visit to the dam site banished doubts and left him in awe of the Grand Coulee undertaking. "I do not believe there is a development in any other part of the world which compares to the Grand Coulee job," he told the *Seattle Times*. The plant, already the source of contracts for Seattle firms, would power mineral mining when completed, he declared. There was, however, bad news for irrigation. In 1937 the Washington legislature refused to continue funding the state's Columbia Basin Survey Commission, created in 1919 to study the problem of irrigating the Columbia basin. The legislature revived the commission in 1943 but the federal government continued to hold the development reins. Constructing the dam, filling the Grand Coulee, World War II, and building the canals delayed the first irrigation water in the Columbia basin until the spring of 1948.[56]

Curtis's interest in the Columbia River was larger than the basin. As the chairman of the Columbia Gorge Committee of the WSPC he was vitally concerned with riverine development from the Columbia's bend below Kennewick through its westerly trend to the Pacific. He and the members of his committee wanted a joint Washington-Oregon commission with power to control development along the river before industrial growth overwhelmed any part of it. As Curtis argued, "while I favor every reasonable development of the Columbia area, the states of Oregon and Washington should not do anything that will destroy the Gorge."

Late in 1938 he reported to a general meeting of the planning council that industry should be barred "within the confined areas of the gorge where it will in any degree destroy its beauties." Curtis and his committee pressed the Washington legislature for a joint committee of state senators and representatives to cooperate with a similar group named by the Oregon legislature. The joint committee of legislators would recommend land use controls to preserve natural beauty, prevent "stream pollution," and channel industrial, recreational, and residential developments into their proper locations. The proposed resolution died in committee. In February 1941 the Washington legislature at last passed a resolution favoring the interstate group. Its first report did not appear until December 1942, a year into the war.[57]

Curtis's committee work did not go unnoticed. He probably never knew why two of his modest rail transportation vouchers of four dollars each were authorized. "I am quite clear these ought to be paid," the council chairman advised its executive officer, "since he [Curtis] is so useful to the Council."[58] The memo was an affirmation of his value because it was private, giving support to Curtis in a small matter. It was no public testimonial designed to shed at least some favorable light on the giver as well as the receiver.

In other ways Curtis affirmed his usefulness to the council. As a member of the Public Works Committee he presented to a 1934 meeting a construction program for state primary and secondary highways, a proposal probably duplicating those of the WSGRA. He was instructed to meet with Lacey V. Morrow, the state director of highways, and thresh out a program to be submitted to the committee. He helped to advance the meeting by moving to pass resolutions and by supporting a Grand Coulee dam high enough for power production and effective irrigation.[59] Although not a member of the council's Olympic Peninsula Committee, he influenced its conservative report on the proposed national park, and its objection to the vast park ultimately created.[60] His was not the only persuasive voice, so his exact impact cannot be known. The council's original proposal for an Olympic park and its subsequent concerns about its size, however, closely paralleled his own.

In all of his reclamation work Curtis adhered to certain principles, though they were never so fixed in his mind as to stand in the way of furthering irrigated farming. Above all else he believed in private property and private enterprise. No enthusiast for federal intervention, he nonetheless recognized that only the national government had the capital and other resources to sustain a national reclamation program. Only the federal government could develop and enforce regulations preventing speculation on irrigable land. Of course a landowner could refuse to sell his property if he believed that the federal valuation of his irrigable acres was too low. He could work to raise the valuation, or he could find some alternate use for his land, a dubious proposition in eastern Washington's scrub desert.

Other governments weren't sidelined. Any effective water use plan for the Columbia River, whether for irrigation, power production, or flood control, should involve state governments in cooperation. When a gravity plan for Columbia basin irrigation was still a possibility, Curtis hoped for a unified water use arrangement involving Montana, Idaho, and Washington. When developing the lower Columbia was at stake, he pressed for a Washington-Oregon compact controlling land use to preserve scenic beauty and a clean waterway while allowing suitable locations for industry, farming, commerce, and family homes. In each town or city along the Columbia's course there was plenty for local governments to do in their traditional areas of responsibility.

These views were hardly those of a political and social troglodyte, but Curtis's forward thinking did not stop with them. He was an enthusiastic supporter of all sorts of cooperative marketing schemes, even though their success was mixed during his ownership of the ranch. For him, farmland ownership implied neighborliness and mutual aid, not a retreat into an imaginary, unachievable independence. Overall, his embrace of reclamation involved a congruent approach, with each entity, public or private, performing in a sphere best suited to its ability and authority.

Chapter Twelve

Working and Living: The Later Years

History is untidy, as many have remarked, and personal lives are no less ragged. Curtis's life never exhibited a single trajectory. He savored some successes after 1935, and surely stayed busy until his death in 1941. Thus it is invalid to argue for his coasting, resting, or simply living out the balance of his life. Yet his great work, becoming a renowned photographer, laboring to develop Rainier, and boosting the scenic and natural resources of his state and the Pacific Northwest, was mature by the mid-1930s.

At the onset of the 1930s the Great Depression dealt him a triple blow. Unmet obligations swamped his ranch until it was foreclosed. The election of 1932 was a disaster for him because the Democrats swept the country. Franklin D. Roosevelt succeeded Republican Herbert Hoover in the White House, a man whom he revered losing to a man on whom he wasted little regard. His daughter Betty remembered a story he told after returning from his 1936 trip to Washington, DC, to testify against a large Olympic national park. One evening he and F. W. "Matt" Mathias of the Aberdeen, Washington, Chamber of Commerce, also a large park opponent, were strolling on a District of Columbia street. A Secret Service agent approached them, offering to take them across the street to meet Roosevelt. The offer evokes a more innocent era of presidential security, but it had an immediate meaning for Curtis. "Dad said never mind," Betty recalled. He "didn't think much of Roosevelt before he was president and hadn't seen anything to change his mind."[1]

The loss of Washington State to the Democrats was its own deeply disturbing development within the national disaster. Curtis consoled his friend and defeated gubernatorial candidate John A. Gellatly with praise for his "good clean, constructive type of fight" against the "cataclysm" of Democratic victory. "Apparently," he wrote in partisan sympathy to a rare Republican survivor, "whatever was loaded into the garbage can and labeled Democratic, was elected."[2] The third blow fell on the Asahel

Curtis Photo Co. The business survived but never recovered from the diminished income of the early thirties.

Curtis's triumphs in the thirties included photo murals of Washington State for the Century of Progress exposition in Chicago (1933-1934), and the San Francisco and New York World's Fairs (1939-1940). More unconventional and perhaps more diverting was his 1931 visit to British Columbia to give an illustrated talk on the beauties of Washington to the king and queen of Siam (now Thailand). Local recognition continued with displays of his world's fair murals and judging of photo contests. A controversial Federal Writers Project guide to Washington published after his death included some "excellent" photos from his collection.[3]

In February 1933 *National Geographic* magazine published a lavishly illustrated article, "Washington, the Evergreen State," as part of a series on the states.[4] The article attributed 32 of the 77 photographs to Curtis, which should have been a great satisfaction to him. It was not. Two photos were incorrectly credited to him, through no fault of his or the magazine's.[5] Curtis worked on the assignment for more than a year, receiving criticism for the exposure, development, and composition of some of his color photographs, which were "not satisfactory either in quality, number or subject." The critics were, however, satisfied with his black and white photos. *National Geographic* sent its own photographer to Washington to record color scenes and to achieve a pictorial balance between eastern and western Washington. Curtis probably lost more money on the transaction than he intended, even though he was willing to absorb part of his costs in return for their booster value.[6]

In fairness to *National Geographic*, the magazine had high standards, and it wished, understandably, to present its readers with a geographically balanced article. The magazine was not alone in criticizing Curtis, because others of his clients were not always easily pleased.[7] Yet the strictures of its staff carry a tinge of East Coast superiority over a distant provincial. In the early 1930s color photography was complex and difficult to handle successfully.[8] The color photos of Clifton Adams, the *National Geographic* photographer, are not superior to Curtis's. The magazine could have found other talented photographers in Washington instead of sending Adams across the continent and around the state. Its editors selected 30 of Curtis's photos from the 69 color and 380 monochromes in his submissions, some of which he had solicited from others. He did not approve of all the choices. "They sent back what I thought were some of the best…and used some that wasn't [sic] worth the space."[9]

One of Curtis's last and largest studio projects was a twenty-volume pictorial history of Washington and the Pacific Northwest, bound in leather. A moneyed widow, Mary S. de Steiguier, asked Curtis to prepare it. He used old photographs in his collection, historic photos purchased by de Steiguier, or contemporary photos taken by him or his staff. The photographs were mounted on quality paper with extensive explanations in Zerby Strong's meticulous lettering. Curtis "was never under contract," Betty recalled, and while both de Steiguier and Curtis "enjoyed working on" the project "very much," their collaboration "was a very formal and precise business relationship."[10] In 1938 de Steiguier transferred all rights in the volumes to Curtis, yielding control of a large, if not especially lucrative, undertaking.[11]

The office staff shrank in the 1930s. Harriet Holmberg left in 1932. "I *released* myself," she wrote fifty years later. She knew "that he couldn't afford me" so she used Betty's need for work as an excuse for her to leave and Betty to take her place. "Holmberg, if you can find something else, it sure would help," Curtis replied. Lindsley was laid off in December 1930, worked sporadically in 1931, but only once in a while thereafter. Others, including Cornelius V. Banta, a skilled photographer and organizer, Strong, Betty, and Eva, stayed on.[12] When Curtis died his workforce was scarcely larger than when he formally took over the Romans studio in 1920. The shrinkage in commissions and employment resulted from the Great Depression, not from any lack of zeal. For more than two decades the Asahel Curtis Photo Co. did what Curtis intended it to do, provide a living for his family and finance his other interests.

The good roads movement continued to be a major interest, but by the time of Curtis's presidency of the WSGRA (1933-1934) road building was almost completely professionalized. From the advent of the Democratic New Deal in 1933 key elements of highway planning and construction became federalized. The close cooperation among state highway departments, the AASHO, and the BPR encouraged centralized government decisions and discouraged private initiatives such as lobbying for this or that improvement on the state level, a principal function of the WSGRA. In 1933 Governor Martin appointed Lacey V. Murrow the director of highways. Murrow, a talented, no nonsense engineer, was as apolitical as anyone in his position could be. He set his own priorities and relied on his staff for road building advice.[13] New Deal agencies, the Civil Works Administration, the Public Works Administration (PWA), and the Works Progress Administration, undertook their own highway improvements. Their activities advanced the interests of the WSGRA but were not subject to its blandishments.[14]

Bureaucratic professionalism was in the saddle to stay, a federal official told the WSGRA in 1938. L. I. Hewes, the regional director of the BPR, was careful to praise the WSGRA, but his compliments were limited to the association's work during the "transition from the mud and dust to smooth pavement" a transition "so pleasant and profitable that the principal grief was to decide who came first with his proposed improvement. But, gentlemen," Hewes intoned, "we have passed the days of the chicken dinner type of allocation of funds." Main highways were improved, but parts of the system were "obsolescent," and cities and secondary road users were demanding more money. The situation required informed choices, not "the indiscriminate allocation of funds under pressure from the marching clubs, even those that probably in the past made real contribution by their support." Despite the diminished role of the "marching clubs," the WSGRA still had a role to play. As it was "a responsible body of citizens in private and public life," it was "capable of intelligent guidance of the public mind" in favor of continued highway development. That development, however, would be determined in administrative and engineering offices, not by pressuring legislators or highway officials.[15]

Curtis and his organization were, to put Hewes's admonition in other terms, the victims of their success. The WSGRA began as a group of road enthusiasts who worked to persuade laggard governments of the central role of roads in the economy and society. The WSGRA struggled so convincingly for a publicly funded, professionally trained, nonpolitical state highway cadre that it finally got what it wanted. In this sense, Curtis was a progressive reformer who shared with others in the late nineteenth and early twentieth centuries an intense effort to expand the role of government as a major player in the economy and the principal guardian of citizen welfare. He did not refer to himself in that way. He would have rejected the label and scoffed at any suggestion that he shared the values of progressives such as Jane Addams or Washington Gladden. He grounded his admiration for political progressives Theodore Roosevelt and Herbert Hoover in their dynamism, their active but sensible living, and their belief in responsible individualism. Their Republican Party affiliation was more important than any policy stance, always excepting conservation. Nevertheless Curtis belongs in the progressive pantheon because of his belief in an energetic, if limited, government.[16]

Curtis's relation to Rainier and other national parks suffered the same fate as his involvement with road building, but with the added bitterness of his estrangement from the Park Service. The probability is that his contribution to Rainier would have decreased anyway. The parks' road program was complete or tentatively decided by the mid 1930s. Any upgrades to intrapark roads would be modest, and would never resemble the freeways or the four-lane streets in cities outside the park. Moreover, since the advent of the New Deal, Curtis's close working relationship with the park superintendent and personal acquaintance with the national parks director were no longer enough to influence the development of Rainier. The New Deal's independent agencies, active in the park, were largely independent of citizen input. The PWA built a fence to screen work buildings at Yakima Park (Sunrise), as well as restrooms at campgrounds and picnic areas there. Elsewhere in the park the PWA built a ranger station, a warehouse, fire lookouts, fire patrol cabins, and other structures while it also improved campgrounds. The work of the Civilian Conservation Corps was even more extensive. The conservation corps built or maintained power and telephone lines, constructed fire breaks and horse trails, planted trees, shrubs, and flowers, built bridges and retaining walls, and erected a variety of buildings, including a community house and a second administration building at Sunrise. Superintendent Tomlinson directed all corps work to be done in cooperation with the park's landscape architect, effectively insulating that official from pressure from Curtis's advisory board or any other outside influence.[17]

If disenchantment over the Olympic park and the break with the Park Service were not enough for Curtis, he was equally distressed over plans for a national park in the North Cascades. From 1938 into 1940 New Deal administrators dusted off an older idea for an extensive national park focusing on the North Cascade Mountains. The conception gained traction within the Park Service, where Tomlinson played an active role in the planning. Opposition within Washington State shelved the scheme, then World War II intervened to focus attention elsewhere. Not until

1968 did the new park, reduced in size from the most ambitious plans, develop.[18] The proposal was just another Interior Department-Park Service land grab, Curtis believed, one designed to lock away vast swaths of timber and mineral resources from business and labor. "Washington needs all of its natural resources to provide work for its people," he wrote in 1939. In 1940 he declared that land should be withdrawn from private development to forestall "exploitation, but not to prevent the use of these resources under government supervision." The "senseless extension of National Parks and monuments" was subverting the idea of careful, controlled development by closing up resources forever.[19]

Bureaucratic instead of private growth was a lesser though linked anathema for Curtis. "He often joked about the fact that if things kept up the whole State of Washington would be one big park and we'd all work for the government," Betty wrote.[20] By then he had embraced the Forest Service as more congenial to his vision of the future, a vision combining controlled resource development with conservation areas, recreational use, and road access.

Curtis's family life altered as his children matured and married. Betty, though married, worked several years for her father, observing him at close range more than her siblings. Her affection for him comes through in her letters to biographer Archie Satterfield. Whitney became the vice president of a Seattle real estate firm. Walter joined the music department of the Seattle-based Cornish School, and continued living in his parents' home, where his wife lived with him after their marriage. Polly, unmarried at the time of her father's death, also pursued a career in music. She wed in 1942. The Asahel-Florence marriage survived its difficulties. Florence "had mental problems all along," Betty related, a situation that "was a great burden" for Curtis. "Work was an outlet and relief for him I'm sure."[21] What Whitney and Walter thought of their father, or he about them and their spouses, is not a subject of reminiscences or of his surviving correspondence. Polly, as mentioned, "resented" her father's detachment. An anonymous interviewer of Betty and Polly in 1982 noted a "tremendous lack of communication" in the family.[22]

On Friday, March 7, 1941, Curtis went to work as usual. It was a better day than normal for Seattle in late winter, with mostly clear skies and a temperature expected to reach the low sixties. About mid-morning Curtis called Betty to say that "he never felt so tired" and wondered "how he'd get through the day," a "most unusual" statement from him. She "urged him to go home," an action he postponed but took later. He thought that he was suffering from indigestion, a misconception common among incipient heart attack victims. He asked Florence to buy some medicine at a drugstore but "he was dead when she returned."[23] He was sixty-seven. His estate was valued at $10,000, a modest sum even in 1941, proof of his disinterest in making money for its own sake.[24]

The morning *Post-Intelligencer* was first with the news of his death, recounting his life on page three of the first section, at the time a significant concession to his wide acquaintance. Both the morning paper and the evening *Times* included a portrait photograph and an outline of his major achievements. Two days later a *P-I* editorial claimed that most outsiders knew of Washington through his photographs. "If he had been no more than a photographer that influence would have been profound. But he was much more." After listing, again, his accomplishments, the editorial found his greatest legacy in his inspiration to others. Washington and the Pacific Northwest needed more people with his "breadth of interest."[25]

A tribute in *Seattle Business*, the magazine of the chamber of commerce, was revealing and remarkably lengthy for a small format, four-page publication. As would be expected, the article began with his service to the chamber. "No one person has given more time or energy to the work of the Chamber or has pursued with greater conscientiousness or diligence its many projects aimed at a greater development" of the state, noting especially his committee service and his long tenure on the board of trustees. After cataloguing his contributions to highways and Rainier, the writer offered an insight. "To tell of all his interests and accomplishments, on many of which he was successful, and some of which he was not, would take many pages." The "not" category would include Rainier's West Side Road and the battle against the Olympic park. The remembrance in the *Washington Motorist* by his friend John P. Hartman recalled his more than thirty years of "virile and active" service to the WSGRA. Curtis was, Hartman wrote, "an artist of rare attainment" and "a lover of nature" whose death was an "irreparable loss" to the WSGRA and Washington State. Hartman honored Curtis again when he delivered a eulogy before the WSGRA annual convention in September.[26]

Other organizations besides the several already noted claimed Curtis's time and attention. He was active in a citizen group, the Seattle Traffic and Safety Council, during the 1930s. There is some evidence that he was accident prone, a condition that may have sparked his interest in traffic safety and may have been partly responsible for his refusal to drive. Among other accidents, he "suffered minor injuries" in July 1937 when an auto struck him as he was leaving a streetcar. He headed a 1938 committee in charge of groundbreaking for the first pontoon bridge across Lake Washington, east of Seattle, and was a member of the 1940 Lake Washington Bridge Dedication Committee responsible for the formal opening of the completed structure. He belonged to the local lodge of the Elks, a national service and social organization.[27]

Florence outlived her husband by more than thirty-three years, though at first her prospects were not so positive. "Mrs. Curtis is in a mental institution (temporarily)," wrote a sympathetic onlooker involved in the effort to salvage the photographs, negatives, and other materials of the Asahel Curtis Photo Co. His October 1941 memorandum mentioned Florence's situation in the course of demonstrating why "the family is in bad shape." About early 1944 Florence moved to Hawaii to live with her younger daughter Polly, and remained there until her death on March 24, 1974. She was eighty-nine.[28]

The Asahel Curtis Photo Co. "is on the rocks," the onlooker continued. His assessment was accurate considering the company's income for 1940, an all-time low of $6,829.20. There is no apparent reason for the scanty receipts during a rebounding economy, just as there is no obvious explanation for their climb to $8,390.96 in 1941, considering Curtis's death in March of that year. The company shut down in 1942, the exact date uncertain. On January 11, 1944, the last check formally closed the business and its bank account.[29] Before that time a movement to preserve the Curtis photographs, negatives, pictorial history, business records, and camera equipment developed within the WSGRA, then spread beyond the road organization. The fund raising testified to Curtis's significance more than any written recognition. After some misdirected effort and an alleged misappropriation rectified by a restitution, the fund reached

$3,000, enough for the Washington State Historical Society (WSHS) to purchase the material in August 1943.[30]

Curtis detested funerals and cemeteries, probably because he associated them with organized religion. The family compromised on a viewing, followed by cremation, but no burial. "Dad hated cemeteries," Betty related, "and informed me many times that he [would] come back to haunt me if I put him in one," so the urn containing his ashes rested in limbo at the funeral home for five years. In 1942 the Forest Service converted an existing camp into a memorial Asahel Curtis Forest Camp of two acres in the Snoqualmie National Forest about fifty miles east of Seattle, south and west of the Snoqualmie Pass summit. There, in 1946, the service placed the urn containing his ashes beneath a modest rock crypt. An attached bronze plaque contained his life dates. It praised his "making known the beauty and scenic wonders of the Pacific Northwest," concluding with "To his memory this bit of the great out-of-doors that he loved so well is dedicated forever."[31]

Then two events prompted a reappraisal of the campground. The first destroyed the Curtis crypt. On October 12, 1962, the regionally famous "Columbus Day Storm" brought winds of eighty-three to ninety-two miles per hour to Washington. Wind toppled a huge fir in the Curtis campground. It fell on the crypt, demolishing it and the urn, scattering Curtis's ashes and ripping the plaque from the crypt. The tree gashed and deformed the plaque, preventing its remounting.[32] The family preferred to believe that a lightning bolt destroyed the crypt even though electrical storms were uncommon in the area. The lightning strike story has achieved legendary status through the repetition of an imagined drama.[33]

The second event was related to the first, though its effect was more gradual if potentially more devastating. Campers and picnickers increasingly availed themselves of the Asahel Curtis Forest Camp, their cars and foot traffic compressing the soil and weakening the trees. Tree loss through windthrow prompted the Forest Service to relocate the campsite and rededicate the original reserve as the Asahel Curtis Memorial Grove. To preserve the area, an Asahel Curtis Nature Trail confined visitors to the trail, while the Forest Service rehabilitated the ancient trees in the grove and supervised its replanting with native species. The service expanded the memorial into the 158-acre Asahel Curtis Recreation Area, provided with camping and picnicking sites, and with trails.[34]

On September 26, 1964, the Forest Service and the East Lake Washington District of the Washington State Federation of Garden Clubs dedicated the nature trail. The event was elaborate and well publicized. Boy Scouts, Girl Scouts, and Campfire Girls participated in the ceremony, as they had in refurbishing the trail and the grove. The speeches focused on the joint work of the Forest Service and the East Lake Washington District in restoring the native habitat of grove and trail, while mentions of Curtis himself were incidental. Betty, Polly, and Whitney, who had begun calling himself Asahel Curtis Jr., were among the estimated 500 who attended.[35]

By the time of the dedication the outpouring of books and heavily illustrated articles about Curtis was underway. Serious scholars, popular historians, and others mined his papers at the University of Washington,[36] as well as related collections, to analyze the development of Mount Rainier National Park, other national and state parks in Washington, or to create illustrated biographies.[37] Some writers described him as a reportorial, artless photographer while others were more generous. In 1979 the National Endowment for the Humanities (NEH) awarded the WSHS an $83,000 grant to develop more prints from the negatives in the Curtis collection, and to prepare at least one traveling exhibit. With NEH and other funds the staff printed a large number of the 40,000 to 60,000 negatives in the collection, bringing the total to 40 or 50 percent by 1983. Interviews preserved personal memories of Curtis, while lectures increased the awareness of his contributions.[38] Another welcome result of this funding was the WSHS's publication of Frederick and Engerman, *Asahel Curtis*, seventy-two pages and paperbound, with lavish use of Curtis photos. The authors selected the illustrations to exhibit a cross-section of Curtis's work, providing thoughtful introductions to topics such as the good roads movement, agriculture, and Neah Bay native whaling. They wrote a brief but positive, sympathetic review of his life and achievements.[39]

Curtis's detractors caricatured him as a creature of resource exploiters, or else as someone so enthralled with economic development that he did not have to be bought. In truth, Curtis wished for, and worked for, economic growth. He believed in businesspeople who could mobilize land, labor, and capital to advance economic development. An expanding economy meant, ultimately, jobs and the stable, prosperous life that only jobs could bring. Though a partisan Republican, he worked with Democrats to advance road construction. He got along well with Democratic Governor Clarence Martin. He bemoaned the Great Depression because it brought Democratic political victories but even more he deplored the loss of jobs. He was no political reactionary, as his advocacy of land use controls for the Columbia River revealed. He tried to balance needs for the wise use of natural resources with the equally demanding desires for recreation in natural surroundings. It was a balance difficult if not impossible to reach, but well worth the struggle to achieve.

Author's Note

The Asahel Curtis material at the University of Washington Libraries' Special Collections is the principal source for the study of Curtis's life. Guides to the Curtis papers and photographs are available online. The Curtis correspondence and clippings are, for all their apparent fullness, neither complete nor adequate. A Curtis diary (1898-1899) covers some of the later part of his stay in Alaska, British Columbia, and the Yukon during the gold rush. It is possible that he kept other diaries, now lost. The correspondence does not begin until 1908, when Curtis was in his thirties, and is spotty until the late 1910s and 1920s. While overflowing on some topics, the papers are scant concerning others, such as the sale of his farm near Grandview, Washington. His daughters, Betty and Polly, "went through his papers after his death...and anything we felt to be of value, and not strictly personal was given to the University of Washington." Betty's comment to biographer Archie Satterfield did not explain the fate of the correspondence deemed to be of no "value" and "strictly personal." Satterfield did not ask.

A few letters in the Curtis papers refer to correspondence missing from the collection. Other letters were folded, then unfolded for inclusion in the files, indicating that Curtis carried them with him for a time. Curtis had his mind on matters besides his correspondence files. Probably he mislaid some letters or withheld them from the files, letters that may be presumed lost. Other collections, such as those in the Washington State Archives, Olympia, contain letters to and from Curtis not found in his papers. Those collections and other pertinent sources, as well as sources once available but now lost, are cited or discussed in the text, bibliography, and notes. The existing sources help to flesh out the Curtis correspondence files at the University of Washington, but lacunae remain.

Principal Abbreviations in the Notes

ACC	Asahel Curtis Collection (WSHS)
ACP	Asahel Curtis Papers, General Correspondence (UW)
ACP2	Asahel Curtis Papers, Mount Rainier Scrapbooks (UW)
ACP3	Asahel Curtis Papers, Speeches and Writings (UW)
AOMS	Scrapbook Relating to Aberdeen and the Olympic Mountains (ACP, UW)
APRCA	Alaska and Polar Regions Collections and Archives (UAF)
AS	Archie Satterfield Collection (APRCA, UAF)
CBR	Curtis Business Records (WSHS)
CMP	Clarence D. Martin Papers (WSA)
CPNS	Center for Pacific Northwest Studies, Bellingham, WA
DC	Digital Collections (UW)
DHSF	(Washington) Department of Highways Subject File (WSA)
EMP	Eula Lee Merrill Papers (UW)
ESMP	Edmond S. Meany Papers (UW)
HD	(Washington) Department of Highways (WSA)
HSF	Gov. Hartley Subject Files (WSA)
LLP	Lawrence Denny Lindsley Papers (UW)
LP	Ernest Lister Papers (WSA)
MAD	Minutes of the Board of Trustees, The Mountaineers (UW)
MRS	Mount Rainier Scrapbooks (ACP, UW)
NARA	Pacific Alaska Region, National Archives and Records Administration, Seattle
P-I	*Seattle Post-Intelligencer*
PMP	Preston P. Macy Papers (UW)
RG 79	Record Group 79, Records of Mount Rainier National Park (NARA)
SC	Special Collections (WSHS)
SVID	Records of the Sunnyside Valley Irrigation District, Sunnyside, WA
Times	*Seattle Daily Times, Seattle Sunday Times*
UAF	Elmer E. Rasmuson Library, University of Alaska Fairbanks
UW	Special Collections, University of Washington Libraries, Seattle
WSA	Washington State Archives, Olympia
WSCPRB	Washington State Council for the Protection of Roadside Beauty
WSDOT	Washington State Department of Transportation
WSGRA	Washington State Good Roads Association Records (CPNS)
WSHS	Research Center, Washington State Historical Society, Tacoma
WSPC	Washington State Planning Council

Notes

Note to the Introduction

1. The fact of an able-bodied male who was a president of the Washington State Good Roads Association not driving a car is so bizarre that it requires some explanation. Sally MacDonald, in "Image Conscious," *Pacific* [magazine of the *Seattle Times* and the *Seattle Post-Intelligencer*], June 12, 1988, 16, indirectly attributes Curtis's refusal to drive to his daughter Polly. Polly's older sister Betty also participated in the widely-ranging interview. She did not contradict her sister. Other support for Curtis's not driving is in a memorandum from B. H. Kizer, chairman of the Washington State Planning Council (WSPC) to P. Hetherton, May 1, 1940, insisting that two railroad travel vouchers from Curtis be reimbursed, box 89, file 4, Records of the Washington State Planning Council, Washington State Archives, Olympia (WSA). A good highway, which Curtis could have driven had he wished to, existed between Seattle and Olympia. A letter from Curtis to Ross K. Tiffany, August 18, 1938, states that he will take a bus from Seattle to Olympia, but needs a ride from Olympia to Portland for a meeting of the Columbia Gorge Committee, box 89, file 1, WSPC, WSA.
2. *Times*, June 20, 1928.

Notes to Chapter One

1. Sarah H. Gordon, *Passage to Union: How the Railroads Transformed American Life, 1829–1929* (Chicago: Ivan R. Dee, 1996), 155, 160–61, 183, 224–25, 233; and John H. White, *The American Railroad Passenger Car* (Baltimore: Johns Hopkins University Press, 1978), 469–72.
2. For the first quotation, see Ann Makepeace, *Edward S. Curtis: Coming to Light* (Washington, DC: National Geographic, 2001), 24. For the second quotation, see Barbara A. Davis, *Edward S. Curtis: The Life and Times of a Shadow Catcher* (San Francisco: Chronicle Books, 1985), 18. The first interview of Eva was by Ralph W. Andrews, *Curtis' Western Indians* (Seattle: Superior, 1962), 18.
3. For the Northern Pacific branch, see Kurt E. Armbruster, *Orphan Road: The Railroad Comes to Seattle, 1853–1911* (Pullman: Washington State University Press, 1999), 85–119. For the mosquito fleet, see Clarence B. Bagley, *History of Seattle: From the Earliest Settlement to the Present Time* (Chicago: S. J. Clarke), 1: 100–21.
4. Information on Johnson Curtis is from *Regimental Muster and Descriptive Roll of Company D, 28th Regiment,* Wisconsin Office of the Adjutant General, State Historical Society of Wisconsin; *Twenty–Eighth Wisconsin Volunteer Infantry, "Whitewater Company No.3," Company D*, www.28thwisconsin.com/rosters/companyd.html; and *Regimental History*, www.28thwisconsin.com/dates28.html.
5. Sok Chul Hong, "The Burden of Early Exposure to Malaria in the United States, 1850–1860: Malnutrition and Immune Disorders," NIH Public Access, www.ncbi.nlm.nih.gov/pmc/articles/PMC2600412.
6. *Regimental Muster of Company D.*
7. *Regimental History*, 2; and "Battle of Helena," *The Encyclopedia of Arkansas History and Culture*, encyclopediaofarkansas.net/encyclopedia/entry-detail.aspx?search=1&entryID=1124.
8. An example of the chaplaincy claim is Davis, *Edward Curtis*, 17. For more on chaplains, see National Civil War Chaplains Museum, chaplainsmuseum.org/production. The fullest account of Johnson's war service in Edward Curtis biographies is Laurie Lawlor, *Shadow Catcher: The Life and Work of Edward S. Curtis*, reprint ed. (New York: Walker,

1994: Lincoln: University of Nebraska Press, 2005), 9–10. Citations are to the University of Nebraska Press edition.

9. Lawlor, *Shadow Catcher*, 10–21.
10. Betty McCullough to Archie Satterfield, October 25, [1969], box 3, file Curtis Mountains, Archie Satterfield Collection, accession 97–166, (AS), Alaska and Polar Regions Collections and Archives (APRCA), Elmer E. Rasmuson Library, University of Alaska Fairbanks (UAF).
11. An early, unvarnished account of the fire is in Bagley, *History of Seattle*, 1: 419–28.
12. A lively account of Seattle's development is Roger Sale, *Seattle: Past to Present* (Seattle: University of Washington Press, 1976), 32–86.
13. Lawlor, *Shadow Catcher*, 21–23, 25, 28–31, and 33–39 for Edward Curtis's life through the Harriman Expedition. For the expedition, see William H. Goetzmann and Kay Sloan, *Looking Far North: The Harriman Expedition to Alaska, 1899* (Princeton: Princeton University Press, 1982. Paperbound ed., 1983).
14. The R. L. Polk Company published Seattle's and many other cities' directories under varying titles. All Seattle directories were listed as published in Seattle. They will be cited as *Polk Directory*, with the date and relevant pages following. The Edward Curtis 1893 listing is on 330. Subsequent years and pages are 1895–96, 254; 1897, 244; 1898, 294; and 1899, 304.
15. This paragraph and the three following are based on Beaumont Newhall, *The History of Photography* (New York: Museum of Modern Art, 1982), 56–279; and the experiences related to me by Katharine L. Wilson, my wife, an amateur photographer and photo developer.
16. Some details of the trip may be recovered from Stephanie Lile, "Two Views, Two Voices: The Stereoscopic Perspective of Photographers Asahel and Edward Curtis," *Columbia: The Magazine of Northwest History* 10 (Spring 1996): 17–28, quotation, 18 (hereafter *Columbia*); the Asahel Curtis *Diary*, which covers from November 1898 to February 1899, box 12, file 16, Asahel Curtis Papers, accession 4058–4 (ACP), Special Collections, University of Washington Libraries (UW); and a last surviving photograph of Dawson taken in the late spring or summer of 1899, no. 46129, Asahel Curtis Collection (ACC), Research Center, Washington State Historical Society, Tacoma (WSHS).
17. *Seattle Argus* (*Argus*), September 18, 1897. The Yukon Archives Genealogy Research Database is available at www.yukongenealogy.com.
18. Davis, *Edward Curtis*, 23.
19. The people dependent on Edward are listed in the 1900 manuscript census, the Twelfth Census of the United States, T623, roll T23 1745, enumeration district 98, p. 64, accessed through the Pacific Alaska Region, National Archives and Records Administration, Seattle (NARA). Curtis, *Diary*, box 12, file 16, ACP, UW.
20. Distances are a matter of some guesswork, depending on a variety of circumstances. The best overall narrative of the Yukon gold rush remains Pierre Berton, *The Klondike Fever: The Life and Times of the Last Great Gold Rush* (New York: Knopf, 1958). A revised edition was published in 1972 as *Klondike: The Last Great Gold Rush, 1896–1899* (Toronto: McClelland and Stewart). See also Archie Satterfield, *Klondike Park: From Seattle to Dawson City* (Golden, CO: Fulcrum, 1993), 46, 135, 147, 151. Experiences in the Klondike region are a favorite topic of potboiler authors, but a good, realistic, popular account is Charlotte Gray, *Gold Diggers: Striking It Rich in the Klondike* (Berkeley, CA: Counterpoint, 2010).
21. Milton A. Dalby, *The Sea Saga of Dynamite Johnny O'Brien* (Seattle: Lowman & Hanford, 1933), 37, 41, 182–86, 219–30, and Asahel Curtis's unpublished article, see n. 22. For the *Rosalie*, see Berton, *Klondike Fever*, 137.

22. A few direct quotations from an unpublished Curtis article containing his Alaska reminiscences survive in Lile, "Two Views," as well as summaries of his narrative.
23. Ibid. Curtis apparently submitted the original typescript, with photographs, to the *Alaska Sportsman* magazine in 1940. Presumably, the magazine purchased the typescript but did not publish it. What probably was a carbon copy of the original remained in the WSHS for many years, where several authors consulted and used it, sometimes imaginatively. The copy is now lost. Seven Asahel Curtis photographs of trail life and abandoned horses were published in Edward S. Curtis, "The Rush to the Klondike over the Mountain Passes," *The Century* 55 March 1898: 692–97, photographs for which Edward took the credit. Photographs of Skagway, trail life, and miners at work appear in several publications, including Richard Frederick and Jeanne Engerman, *Asahel Curtis: Photographs of the Great Northwest* (Tacoma: Washington State Historical Society, 1983, reprinted in 1985 and 1986), 12–13, 15–19.
24. William R. Hunt, *North of 53°: The Wild Days of the Alaska-Yukon Mining Frontier, 1870–1914* (New York: Macmillan, 1974), 47.
25. Other narratives of Curtis's gold rush experiences, with photographs, include William James Betts, "Klondike Photographer," *Alaska Sportsman* 30 (December 1964): 16–19; Don Duncan, "Curtis, Asahel (1874–1941), Photographer," essay 8780, *HistoryLink.org*, www.historylink.org/index.cfm?DisplayPage=output.cfm&file_id=8780, October 28, 2008, 2–4; and Richard Frederick, "Asahel Curtis and the Klondike Stampede," *Alaska Journal* 13 (Spring 1983): 113–21. A map of the trail is in Satterfield, *Klondike Park*, 134, 148, 156.
26. Russell A. Bankston, *The Klondike Nugget* (Caldwell, ID: Claxton, 1935), 15–46, quotation, 44.
27. Ibid., 48–57.
28. Yukon Archives Genealogy Research, n. 17.
29. Kathryn Winslow, *Big Pan-Out* (New York: Norton, 1951), 185.
30. Readers who wish more information about Klondike mining should read Murray Morgan, *One Man's Gold Rush: A Klondike Album* (Seattle: University of Washington Press, 1967), 132–38; diagrams, 136–37, and excellent maps, 140 and endpapers.
31. Ibid.
32. Curtis, *Diary*, box 12, file 16, ACP, UW.
33. The consolidation of claims and shift to corporate methods of gold extraction is described in Lewis Green, *The Gold Hustlers* (Anchorage: Alaska Northwest, 1977). The clock is pictured in Curtis 985 (Edward's system) UW order no. CUR 1966, Digital Collections (DC), UW.
34. All credit for uncovering this letter goes to Mick Gidley, *Edward S. Curtis and the North American Indian, Incorporated* (Cambridge, England: Cambridge University Press, 1988: paperbound ed., 2000). Citations below are to the paperbound edition. Gidley, however, draws the mistaken conclusion that Edward and Asahel went to the Klondike together in 1897, an assertion repeated in Makepeace, *Edward Curtis.* See Gidley, 55–56; Makepeace, 31–33; and Joanna Cohan Scherer, *Edward Sheriff Curtis* (London: Phaidon, 2008), 6. Timothy Egan, Edward's most recent biographer, has the older brother "dispatching" Asahel. Then "Curtis followed [Asahel] shortly thereafter." See Timothy Egan, *Short Nights of the Shadow Catcher: The Epic Life and Immortal Photographs of Edward Curtis* (Boston: Houghton Mifflin Harcourt, 2012), 30. Edward did not reach Skagway and the White Pass until 1899, when the Harriman Expedition, of which he was the official photographer, spent two days there during June. The party traveled by rail to the summit, tarried a while, then returned to Skagway. Goetzmann and Sloan, *Looking Far North,* 55–67.
35. Edward Curtis, "Rush to the Klondike," quotation, 697.

36. A brief summary of both sides of the Asahel–Edward split is in Alexander I. Olson, "Heritage Schemes: The Curtis Brothers and the Indian Moment of Northwest Boosterism," *Western Historical Quarterly* 40 (Summer 2009): 163–64. The estrangement is recorded in almost every biography of Edward Curtis. See, for examples, Davis, *Edward Curtis,* 31–32; Egan, *Short Nights,* 31; Lawlor, *Shadow Catcher,* 45; and Makepeace, *Edward Curtis,* 33–34, 132. At least five of the seven photos in Edward's article are in ACC, WSHS. They are noted by the ACC photo number, then by the underline in Edward's article, and finally by the page number in the article. 46015, Deserted horses at the foot of "the summit," Skagway Trial, 695. 46032, In camp at Summit Lake on the Skagway Trail, 695. 46110, Near the summit of Porcupine Hill, Skagway Trail, 696. 46121, The Climb to the summit of Chilkoot Pass, Dyea Trail, 692. 46122, On the summit of the Chilkoot Pass, 693.
37. MacDonald, "Image Conscious," 18. Edward's refusal to recognize Asahel's authorship also was in line with the practice, fairly common then, for photographers to use one another's work without acknowledgement of intellectual property rights. Edward's supporters chose not to defend him on those grounds, probably because such unacknowledged borrowing within a family would appear to be in bad taste, however acceptable if conducted across familial boundaries.
38. Davis, *Edward Curtis*, 31.
39. For a description of Edward's household, see manuscript census, n. 19.
40. *Polk Directory*, 1901, 378.
41. "Image Conscious," 18 (first quotation); 34 (second quotation).
42. Asahel Curtis Jr., interview by Alex Olson, May 1981, transcribed May 2, 2006, box Interviews, Special Collections (SC), WSHS.
43. Davis, *Edward Curtis*, 69, 74.
44. *Deeds, Kitsap County, WA,* vol. G, 42–44, 711. On October 17, 1902, Edward and his wife Clara sold the remainder of the farm, presumably 60 acres, for $6,400. Whether any of the proceeds went to Ellen is not noted in the record of the sale, book 36, 14, Puget Sound Regional Archives Branch, Washington State Archives, Bellevue.
45. MacDonald, "Image Conscious," 34, and McCullough to Satterfield, n.d. but 1969 or 1970, box 3, file Curtis Mountains, AS, APRCA, UAF.
46. For the first quotation, see *Times,* November 30, 1902. See also the wedding announcement, box T-151, file Curtis, Asahel, Misc. Letters, Memorabilia (original), SC, WSHS; manuscript census, 1900, roll 6, enumeration district 85, page 208, NARA Seattle (fuller citation n. 19); and *Polk Directory*, 1902, 324. For the second quotation, see Junius Rochester, "Prefontaine, Father Francis Xavier," essay 3633, *HistoryLink.org*. For Florence's later religious affiliation, see *Times,* September 6, 1930.
47. McCullough to Satterfield, September 20, 1969, box 3, file Curtis Mountains, AS, APRCA, UAF. For Asahel's occupations and return, see *Polk Directory*, 1905, 388; 1906, 376; 1907, 366; and 1908, 391. See also Curtis to Romans, October 17, 1908, box 1, file 4, ACP, UW.
48. The professional careers of Asahel, Miller, and Romans are listed in the *Polk Directory*, 1909, 426; 1910, 460; 1914, 597; 1915, 1832; 1917, 576 and 1149. For Romans see also 1912, 2039; and 1916, where he is not listed. For the 1919 quotations, see *Polk Directory*, 121, first quotation; and 1521, second quotation.

Notes to Chapter Two

1. The incorporation is in Ledger 300 18–12, pp. 2, 15, 17, 35, 85, box Asahel Curtis—Articles of Incorporation—Account Books—Ledgers—Loose Inventory Sheets, (box AI), Curtis Business Records (CBR), WSHS. See the *Polk Directory*, 1921, 200; 1922, 544 and

545, and 1925 for the deletion of the Romans name. For the Colman Building, see 1907, 989, and 1941, 398.

2. 1910 manuscript census, roll T624-1658, 78; 1920 manuscript census, roll 7625-1927, 78; and 1930 manuscript census, roll 2497, 84. See also ch. 1, n. 19. The moves of the Curtis family may be followed in the *Polk Directory,* 1905, 388; 1907, 266; 1908, 341; 1913, 573; 1914, 576; 1915, 570; 1916, 563–64; 1917, 576; and 1932, 374.
3. For Eva's addresses through 1917, see n. 2. *Polk Directory*, 1918, 615; 1919, 609; 1920, 6211. The 1930 manuscript census lists Eva, roll 2497, 84; and see ch. 1, n. 19.
4. For the Alaska trip, see Betty McCullough to Archie Satterfield, April 11, [1970]; and for Florence's mental problems, McCullough to Satterfield, October 25, [1969], box 3, file Curtis Mountains, AS, APRCA, UAF. The itinerary and purposes of the trip are in *Times,* June 20, 1913; and June 22, 1913. For the federal railroad legislation and route selection, see William H. Wilson, *Railroad in the Clouds: The Alaska Railroad in the Age of Steam, 1914–1945* (Boulder, CO: Pruett, 1977), 25–30.
5. Information on the Woman's Century Club and its members is on its website, www.womanscenturyclub.org/history. Florence's activities may be followed in *Times,* August 24, 1930; January 26, 1931; and May 1, 1932.
6. For Florence's attending events including Walter, see *Times,* January 13, 1932; and Polly, *Times,* February 18, 1940. She did not limit herself to her children's appearances, see *Times,* February 4, 1939. For the Camp Fire Girls, see *Times,* April 9, 1933.
7. For the Park Service, see Stephen T. Mather to Curtis, March 17, 1916, box 1, file 29; and Curtis to Mather, November 6, 1918, box 1, file 30, ACP, UW. For the Bureau of Public Roads, see J. A. Elliott to Curtis, January 18, 1926, box 3, file 12, ACP, UW. For the Bureau of Reclamation, see Elwood Mead to Curtis, November 1, 1929, box 6, file 27, ACP, UW. For government work, see Curtis to Frank Carpenter, May 16, 1929 (quotation), box 6, file 10; and Curtis to F. O. Hagie, January 27, 1934, box 10, file 23, ACP, UW.
8. For the Northern Pacific, see Curtis to W. H. Paulhamer, August 28, 1922, box 2, file 5, ACP, UW. For the Great Northern, see no. 44095, ACC, WSHS. For the CMS&P (Milwaukee Road), see Curtis to A. L. Eidemiller, October 24, 1929, box 6, file 26 and attachments, ACP, UW. For the Seattle Chamber of Commerce, see no. 25558, ACC, WSHS. For the Spokane Chamber of Commerce, see CBR, box T 314.2-288, WSHS. For the Yakima Chamber of Commerce, see no. 58792, ACC, WSHS. See also Carlos A. Schwantes, "The Milwaukee Road's Pacific Extension, 1909–1929: The Photographs of Asahel Curtis," *Pacific Northwest Quarterly* (*PNQ*) 71 (January 1981): 30–40. National, regional, and local clients are found in the sources cited below.
9. Frederick and Engerman, *Asahel Curtis*; Archie Satterfield, *Seattle: An Asahel Curtis Portfolio* (San Francisco: Chronicle Books, 1985); David Sucher, ed., *The Asahel Curtis Sampler: Photographs of Puget Sound Past* (Seattle: Puget Sound Access, 1973); Asahel Curtis Photo Co. Photographs, 1853–1941, Collection 982, DA, UW; and ACC, WSHS.
10. *Times,* April 21, 1917; and May 21, 1917. Curtis to T. H. Martin, October 13, 1925, box 3, file 2, ACP, UW.
11. For the "know your state" idea, see Curtis to S. B. Nelson, January 5, 1926, box 3, file 10; and for the disorganization of Marsh-Curtis, see Curtis to W. G. Oves, January 22, 1926, box 3, file 13, ACP, UW. For a carbon, see Curtis to G. N. Worden, February 11, 1926, box 3, file 17, ACP, UW. An example of Curtis letterhead slogans is Curtis to W. H. Little, April 26, 1930, box 7, file 22, ACP, UW.
12. One example among many is *Yakima Valley Visitor Guide, 2012* (Yakima, WA: Yakima Valley Newspapers, L.L.C., 2012).
13. Curtis to J. A. McKinnon, April 4, 1927, box 4, file 11, ACP, UW.
14. Curtis to A. C. Clausen, July 19, 1929, box 6, file 24, ACP, UW.

15. Curtis to C. G. Ware, October 18, 1912, box 1, file 17, ACP, UW.
16. For the first two quotations, see Curtis to T. H. Martin, November 21, 1924, box 2, file 27; and for the "untold" quotation, see Curtis to John F. Miller, March 14, 1924, box 2, file 18, ACP, UW.
17. Curtis to E. A. Smith, July 22, 1932, box 9, file 32, ACP, UW.
18. Curtis to T. H. Martin, November 3, 1928, box 5, file 24, ACP, UW.
19. For the incorporation of the Smyser firm, see *Times,* July 12, 1937; and for the contract and other information, see *Times,* August 14, 1938.
20. Sucher, ed., *Asahel Curtis.* The pages of the Morgan and Sucher contributions are not numbered. A Curtis airplane photo is in Satterfield, *Seattle*, 100.
21. Satterfield, unpublished typescript, 84, 85, box 3, file Curtis, AS, APRCA, UAF.
22. Morgan, *One Man's Gold Rush*, 51–57, and, for interior scenes, 159, 161, 164–65.
23. Frank La Roche Photograph Collection 283, DA, UW, nos. 6, 7, and 18.
24. Frank H. Nowell Photograph Collection 316 DA, UW.
25. Essays accompanying their digital collections give biographical information on La Roche and Nowell. For Nowell, see also Nicolette Bromberg, *Picturing the Alaska-Yukon-Pacific Exposition: The Photographs of Frank H. Nowell* (Seattle: University of Washington Press, 2009), 13–23. For Hegg, see Morgan, *One Man's Gold Rush*, 6. Arthur Churchill Warner photographed a wider range of subjects than Curtis, but was more interested in Dyea, Skagway, and the Chilkoot; see Arthur Churchill Warner Collection 273, DA, UW.
26. Easily accessible comparisons between Curtis and Hegg may be made from Frederick and Engerman, *Asahel Curtis*, 12, 15–19; and Morgan, *One Man's Gold Rush*, 94–101, 145–55. See also Satterfield, *Klondike Park*, 41, 116, 118.
27. The "trail" photo is no. 46036, ACC, WSHS; and Frederick and Engerman, *Asahel Curtis*, 15.
28. The Ainsworth photo is no. 46171, ACC, WSHS; and Frederick and Engerman, *Asahel Curtis*, 19.
29. For Warner, see nos. 2, 27, and 48, 273, DC, UW; and for Curtis, nos. 7, 102, 1583, and 1794, DC, UW. Among other Curtis sources, see Satterfield, *Seattle*, 45–46, 49. For Kinsey, see Ralph W. Andrews, *This Was Logging! Selected Photos of Darius Kinsey* (Seattle: Superior, 1954); and the more complete and accurate Dave Bohn and Rodolfo Petscheck, *Kinsey, Photographer: A Half Century of Darius and Tabitha May Kinsey.* 3 vols. (San Francisco: Prism, 1978–1984).
30. Olson, "Heritage Schemes," 159–78. For Gidley, see ch. 1, n. 34.
31. Morgan, Foreword, in Sucher, ed., *Asahel Curtis*, n.p.; and Lile, "Two Views, Two Voices,"17–28.
32. See the defense of Edward in Stedman Upham and Nat Zappia, *The Many Faces of Edward Sherriff Curtis: Portraits and Stories from Native North America* (Tulsa, OK: Gilcrease Museum and Thomas Gilcrease Association; Seattle: University of Washington Press, 2006). See also Gidley, *Curtis*, 109–33.
33. Paige Raibmon, *Authentic Indians: Episodes of Encounter from the Late-Nineteenth-Century Northwest Coast* (Durham: Duke University Press, 2005), 82; and Frederick and Engerman, *Asahel Curtis*, 50–55. For one view of the "authenticity" issue, see Raibmon, 74–97, 133–34.
34. Curtis to B. O. Foss, December 27, 1922, box 2, file 8, ACP, UW.
35. Office staffing is compiled from various sources as noted below.
36. For inflation figures, see www.westegg.com/inflation. The surplus in 1935 was an astounding $2,668.28. A drop of total income of $6,829.20 in 1940 was unexplained, see untitled ledger, box AI, CBR, WSHS. For quotation, see Curtis to Frank Guilbert, November

17, 1933, box 10, file 20, ACP, UW. See also ledgers 1932–1935 and 1936–1940, box 19, ACP, UW.

37. Lindsley, *Diary*, February 27, 1914; February 17, 1916; December 13–15, 1916; January 7, 1917; January 10, 1917 (quotation); November 20, 1917; December 15, 1917; February 2, 1918 ("snake" quotation); February 28, 1923 ("Curtis" quotation); March 1, 1923; May 21, 1927; and December 16, 1927, box 12, Lawrence Denny Lindsley Papers (LLP), 2179, UW. Another analysis of Lindsley and the Lindsley-Curtis relationship is in Chery Kinnick, "Lawrence Denny Lindsley," *Columbia* 27 (Summer 2013): 3–9.
38. Lawrence Denny Lindsley, interview by Harry Majors, 1973, UW. For Edward's fall into obscurity (except perhaps in Seattle), see Davis, *Edward Curtis*, 70. For Strong's return, see Lindsley, *Diary*, April 7, 1920, box 12, LLP, UW. For Asahel's working on Edward's project, see Lindsley, *Diary*, April 18, 1919, box 12, LLP, UW; and for a similar instance, see Davis, *Edward Curtis*, 66.
39. For Lindsley's dislikes, see *Diary*, July 2, 1917; November 5, 1920; March 8, 1924; and December 16, 1927, box 12, LLP, UW. For Lindsley's home studio, see *Diary*, March 10, 1920, box 12, LLP, UW. For Webster & Stevens, see *Diary*, August 19, 1919, box 12, LLP, UW. For Lindsley's marriage, see *Diary*, August 31, 1918; and for his wife's death, June 6–14, 1920, box 12, LLP, UW.
40. MacDonald, "Image Conscious," 15; and Betty McCullough and Polly Kella, interview by anonymous, August 1982, Honolulu, HI, box 314.2, file 28, SC, WSHS. Curtis to Romans, October 17, 1908, box 1, file 4, ACP, UW. Quotations from Betty are in her letter to Satterfield, September 20, 1969, box 3, file Curtis Mountains, AS, APRCA, UAF.
41. Harriet Holmberg Williams, Adele Hubbard, and Una Tinker Eastman, interview by anonymous but probably Richard Frederick and Jeanne Engerman, n.d., box A. Curtis audio tapes, SC, WSHS.
42. Satterfield typescript, 67. The Lindsley interview is lost but quoted in the unpublished typescript, AS, APRCA, UAF. A similar Yukon presentation is noted in *Times,* February 26, 1915.
43. Unattributed paper titled "STUDIO," box 3, file Curtis Documents, AS, APRCA, UAF.
44. For "dour" quotation, see Archie Satterfield to "Dear Charles," February 11, 1977, box 5, unaccessioned location M1468, AS, APRCA, UAF. For Curtis, organized religion, and intrusive women, see McCullough to Satterfield, October 25, [1969], box 3, file Curtis Mountains, AS, APRCA, UAF. For the comments by Curtis's friend William O. Thorniley, see Satterfield's unpublished typescript, 70–72, box 3, file Curtis, AS, APRCA, UAF. Like the Lindsley interview cited in n. 42, the Thorniley interview does not survive but is quoted in Satterfield's unpublished typescript.
45. Thorniley interview by Satterfield, unpublished typescript, 71 (lumber company); and 76–77 (slide show), box 3, file Curtis, AS, APRCA, UAF.
46. *Argus*, November 24, 1931.
47. William H. Wilson, "The War League: Asahel Curtis's Plan for Peace," *Columbia* 23 (Winter 2009–10): 33–35.
48. MacDonald, "Image Conscious," 15–16.
49. McCullough to Satterfield, October 25, [1969], box 3, file Curtis Mountains, AS, APRCA, UAF.
50. McCullough to Satterfield, September 20, 1969, box 3, file Curtis Mountains, AS, APRC, UAF (humor quotations); and MacDonald, "Image Conscious," 15 (Dave Beck).
51. McCullough to Satterfield, September 20, 1969; and October 25, [1969] (quotations), both in box 3, file Curtis Mountains, AS, APRCA, UAF.

52. McCullough to Satterfield, October 25, [1969], box 3, file Curtis Mountains, AS, APRCA, UAF; and Lindsley interview by Satterfield, unpublished typescript, 67, box 3, file Curtis, AS, APRCA, UAF.

Notes to Chapter Three

1. John E. Tuhy, *Sam Hill: The Prince of Castle Nowhere*, reprint ed. (Beaverton, OR: Timber Press, 1983; Maryhill Museum of Art, 1992). Citations are to the Maryhill Museum edition. See also John P. Hartman, *A Brief History of the Washington State Good Roads Association (WSGRA)* (Walla Walla, WA: The Association, 1939), WSDOT Library, Washington State Department of Transportation, wsgrta.com/history_files/A Brief History.pdf.
2. Tuhy, *Sam Hill*, 197–212, for farm community; and 212–17, for concrete mansion.
3. The good roads movement has generated an enormous amount of printed material. For introductions to the movement, see these books and their notes and bibliographies: Stephen B. Goddard, *Getting There: The Epic Struggle between Road and Rail in the American Century* (New York: Basic Books, 1994), 3, 44–45, 54–55, 60–61; Owen D. Gutfreund, *Twentieth-Century Sprawl: Highways and the Reshaping of the American Landscape* (New York: Oxford University Press, 2004), 9–28; Tom Lewis, *Divided Highways: Building the Interstate Highways, Transforming American Life* (New York: Viking Penguin, 1997), 8; Phil Patton, *Open Road: A Celebration of the American Highway* (New York: Simon & Schuster, 1986), 54–61, 145; Bruce E. Seely, *Building the American Highway System: Engineers as Policy Makers* (Philadelphia: Temple University Press, 1987), 11–45; Earl Swift, *The Big Roads: The Untold Story of the Engineers, Visionaries, and Trailblazers Who Created the American Superhighways* (Boston: Houghton Mifflin Harcourt, 2011), 15–45, 66–69, 161; and Christopher W. Wells, *Car Country: An Environmental History* (Seattle: University of Washington Press, 2012), 65–85.
4. Hartman, *Brief History WSGRA*, 4.
5. WSGRA, "Basic Principles of the Washington Good Roads Association" [1928], typescript, box 1, Administrative Records, 3-5-6 (AR), file History 1935–1987 (H), Series I, Washington State Good Roads Association (WSGRA), 1910–1987, Washington State Good Roads Association Records (WSGRA Records), Center for Pacific Northwest Studies, Bellingham, WA (CPNS).
6. For "flivvers" quotation, see Hartman, *Brief History WSGRA*; and for "in addition" quotation, see "Basic Principles," CPNS. See also *Forty Years with the Washington Department of Highways*, 2, WSDOT Library, Washington State Department of Transportation, www.wsdot.wa.gov/NR/rdonlyres/65A1FA26-0800-428A-A2AD-BEBA658D7A4F/0/40yearsReport.pdf; and Tom W. Holman paper, WSGRA, *Proceedings*, September 27–28, 1935, Olympia. The number of the annual meeting was sometimes misstated, so the numbers will be omitted in favor of listing the year only. The titles of the printed annual records of the WSGRA varied. Some printed records were paginated, while others were paginated in part. Because all of the printed annual records were of modest length, pagination will be omitted.
7. For the development of highway legislation, see "The Progressive Highway Construction Program of the State of Washington," typescript, box 65, file Correspondence G 1917, Department of Highways, 1909–1922 (HD), Washington State Archives, Olympia (WSA).
8. Ibid.
9. WSGRA, "Basic Principles," WSGRA Records, CPNS.
10. Tuhy, *Sam Hill*, 130–34; and, for quotations, William H. Wilson, "'Names Joined Together as Our Hearts Are': The Friendship of Samuel Hill and Reginald H. Thomson," *PNQ* 94 (Fall 2003): 191. For the context of Hill's views on rural living, see William L. Bowers,

The Country Life Movement in America, 1900–1920 (Port Washington, NY: 1974), 34–36, 77–79, 112–13.

11. Wilson, "Names Joined Together," 187, 189; and Tuhy, *Sam Hill*, 136–40.
12. Michael J. Berger, *The Devil Wagon in God's Country: The Automobile and Social Change in Rural America, 1893–1929* (Hamden, CT: Archon Books, 1979), 13–39, 88–90.
13. WSGRA, *Proceedings*, September, 29–30, 1933, Port Angeles, and *Proceedings*, September 17–18, 1934, Yakima, contain hints of attempting to win over rural people. For Hill's changing interests, see Tuhy, *Sam Hill*, 137–39.
14. No copy of Hill's letter survives in the archives of the Maryhill Museum, where Hill's papers are housed, or in Hay's governor's papers at WSA. Much of Hill's letter may be deduced from the direct and indirect quotations in Hay's response to Hill, July 1, 1911, box 1, file 23, ACP. For Hill as "Chairman, State Advisory Board," see Bowlby to Hill, January 17, 1910, box 4, file Correspondence, 1910, January–February, HD, WSA.
15. Tuhy, *Sam Hill*, 137–39, quotation, 138.
16. There is no source for the Hill "roads without Hay" quotation. For Hill's taking credit, see Tuhy, *Sam Hill*, 139. For the historian's interpretation of Hay, see Carlos Arnoldo Schwantes, *The Pacific Northwest: An Interpretive History*, rev. and enl. ed. (Lincoln: University of Nebraska Press, 1966), 349.
17. Hay to Curtis, July 3, 1912; and Curtis to Hay, July 8, 1912, both in box 1, file 23, ACP, UW.
18. For Lister's perceived shortcomings and the Columbia River highway, see Tuhy, *Sam Hill*, 139, 147–56. For some of Lister's good roads work, see Lister to Hill, January 6, 1913, and proclamations dated May 13, 1913 and May 2, 1914, box 24-2-58, file Highways–Good Roads, Ernest Lister Papers (LP), WSA. Hill is quoted in E. L. Farnsworth to *Western Motor Car*, January 12, 1916, box 24-2-57, file Highway Commission (1), LP, WSA.
19. Lister to H. J. Snively, March 29, 1916, box 24-2-57, file Highway Commission (1), LP, WSA; and WSGRA, *Proceedings*, September 29–30, 1933, Port Angeles. For Hill's later activity in the good roads movement, see Tuhy, *Sam Hill*, 144–47. For an example of lionizing Hill, see WSGRA, *Proceedings*, November 21–22, 1930, Wenatchee.
20. Hill to Meany, August 23, 1918, box 34, file 12, Edmond S. Meany Papers (ESMP), accession 106-001, UW.
21. Richard F. Weingroff, "Federal Aid Road Act of 1916: Building the Foundation," *Public Roads* 60 (Summer 1966), www.fhwa.dot.gov/publicroads/96summer/p.su2fcm; and Seely, *Building American Highway System*, 46–51.
22. Seely, *Building American Highway System*, 59–65.
23. For federal funding, see Lewis, *Divided Highways*, 18. For state funding, see WSGRA, *Facts and Figures on Highway Financing and Administration in the State of Washington* ([Seattle]: WSGRA, 1931), fig. 5, p. 7. For federal activity in transportation generally, see David W. Jones, *Mass Motorization + Mass Transit* (Bloomington: Indiana University Press, 2008), 11–14, 46–54, 57–94, and Mark H. Rose, Bruce E. Seely, and Paul F. Barrett, *The Best Transportation System in the World* (2006; repr., Philadelphia: University of Pennsylvania Press, 2010), 1–75, 104–05.
24. This discussion of the gasoline tax depends upon John C. Burnham's pioneering article, "The Gasoline Tax and the Automobile Revolution," *Mississippi Valley Historical Review* 48 (December 1961): 435–59. The Washington experience generally supports the excellent analysis and well-documented conclusions of the author. See also Wells, *Car Country*, 185–93.
25. *Forty Years*, 9, 11, 12, 13, 14, 25, 26; and WSGRA, *Proceedings*, September 27–28, 1935, Olympia.

Notes to Chapter Four

1. Albert Francis Gunns, "Roland Hill Hartley and the Politics of Washington State" (master's thesis, University of Washington, 1963), 94 (first quotation), 148 (second quotation); and Curtis to Grant Hinkle, December 28, 1925, box 3, file 5, ACP, UW.
2. Gunns, "Roland Hartley," 187 (quotation), 185, highways, and 223, friendly conversation. For primaries, see 42, 1916; 58, 1920; 68–70, 1924; and, for 1924 general election, 210–16.
3. Curtis to Robert Moran, November 26, 1929, box 6, file 30, ACP, UW.
4. Gunns, "Roland Hartley," 1–7, early years; 7–19, early business and political activity; 39, 52–57, 65–68, later political involvement; and Norman H. Clark, *Mill Town: A Social History of Everett, Washington* (Seattle: University of Washington Press, 1970), 75.
5. Gunns, "Roland Hartley," 73–103, 206–17.
6. Clark, *Mill Town*, 75.
7. Hartley is quoted in *A Statement by the "Majority" of the House of Representatives, Extraordinary Session, Legislature, 1925, State of Washington, 1925* [Olympia?], 6.
8. Gunns, "Roland Hartley," 100–03. See also 112–21, 128–29, 164–67, 173–77, 185–86.
9. Ibid., 101.
10. *Statement by the "Majority,"* 6–7.
11. Curtis to James Allen, January 28, 1925, box 2, file 30; Curtis to Frank W. Guilbert, September 28, 1928, box 5, file 22; and Curtis to Wellington Pegg, November 15, 1930, box 8, file 10 (quotation), all in ACP, UW.
12. Gunns, "Roland Hartley," 190 (quotations about Allen and Hoover); and Curtis to Hartley, February 24, 1925, box 2, file 32, ACP, UW.
13. Curtis to Mrs. Josephine Corliss Preston, April 1, 1926 ("obnoxious" quotation), box 3, file 21; Curtis to Frank Poole, October 2, 1926 (other quotations), box 3, file 35; Curtis to Frank W. Guilbert, August 25, 1926 (highway damage and accidents), box 3, file 32; and Curtis to Orpheus C. Soots, October 12, 1926 (mistaken repeal), box 3, file 36, all in ACP, UW.
14. Curtis to Christy Thomas, January 21, 1927 (quotation), box 4, file 2, ACP, UW. The less important road was a naval torpedo station at Keyport, see A. F. Hamp to Curtis, February 21, 1927, box 4, file 5, ACP, UW.
15. Gunns, "Roland Hartley," 27, 77, n13 p. 77, 88, 150.
16. Ibid., 190.
17. *Times*, March 22, 1927 (first quotation); and March 24, 1927 (second quotation). For Hartley's vetoes and approvals, see *P-I*, March 22, 1927, and March 23, 1927.
18. Gunns, "Roland Hartley," 189.
19. Ibid., 192 (first and third quotations); and *State ex. rel. Clausen v. Hartley*, 257 Pac. 396, p. 140 (second quotation).
20. Gunns, "Roland Hartley," 194–95.
21. For the WSGRA intervention and quotation, see statement by the WSGRA committee to Hartley, Clausen, and Potts, May 12, 1927, and accompanying resolution, box 2K-1-21, file Highway Committee Minutes (1927-2) (HCM), Gov. Hartley Subject Files (HSF), WSA. For Hartley's reply, see box 2K-1-21, HCM, HSF, WSA. For the court decision, see n. 19. For Hartley's concession, see his statement, June 23, 1927; and for his anger, see his statement, October 20, 1927, both in 2K-1-21, file HCM (1927), HSF, WSA.
22. Curtis to Chas. W. Saunders, March 20, 1927, box 4, file 10, ACP, UW. For Bullitt and the Hartley-Bullitt campaign, see Gunns, "Roland Hartley," 208, 215–17.
23. *Forty Years*, 14; and *Facts and Figures on Highway Financing and Administration in the State of Washington*, WSGRA, 1931.
24. *Facts and Figures*, 4 (quotation).

25. Ibid., 8, 9. Curtis to Clark V. Savidge, April 19, 1929, box 6, file 10 ("unbalanced" quotation); and Curtis to Robert Moran, November 12, 1930, box 8, file 10 ("definite hostility" quotation), both in ACP, UW.
26. *Facts and Figures*, 9 (quotation). For similar Curtis comments, see Curtis to Herbert Evison, February 11, 1929, box 6, file 2; and to O. A. Tomlinson, February 30, 1929, box 6, file 4, both in ACP, UW.
27. WSGRA, *Proceedings*, September 11–12, 1931, Centralia; and for the executive committee meeting, see WSGRA, *Proceedings*, September 23–24, 1932, Spokane. For Carlson's comments, see meeting of the North End Improvement Council (NEIC), minutes October 7, 1931 (date omitted hereafter unless necessary for clarity), box 4, file 7, p. 67, Series II, North End Improvement Council & Other Organizations (NEIC), WSGRA Records, CPNS.
28. For Hartley's speech, see WSGRA, *Proceedings*, 1931. For resolutions favoring Humes, see WSGRA, *Proceedings*, September 5–6, Bellingham; and WSGRA, *Proceedings*, November 21–22, Wenatchee.
29. Curtis to Frank W. Gilbert [sic], October 14, 1927, box 4, file 28, ACP, UW. For the popularity of oiling, see WSGRA, *Proceedings*, October 11–12, 1928, Walla Walla; and *Proceedings*, 1929. For Curtis's conversion, see Curtis to Wellington Pegg, November 15, 1930, box 8, file 10; and Curtis to Reed, October 26, 1931, box 9, file 4, both in ACP, UW. For Reed and Hartley, see Robert E. Ficken, *Lumber and Politics: The Career of Mark E. Reed* (Santa Cruz, CA: Forest History Society, Inc., and Seattle: University of Washington Press, 1979), 108–20; and Gunns, "Roland Hartley," 112–21, 128–29, 173–77, 185–86.
30. Gunns, "Roland Hartley, 218–20 (quotations, 219). For Wilmer, see WSGRA, *Proceedings*, 1933. Curtis continued to favor a highway commission, WSGRA, *Proceedings*, September 17–18, 1934, Yakima.
31. *Times,* September 30, 1933.
32. *Seattle's Business* 13 (May 16, 1929): 2; and *Times,* February 22, 1927, January 22, 1928, and September 22, 1940.
33. Curtis to P. H. Sceva, January 23, 1930, box 7, file 5, ACP, UW.
34. "Where the Highway Money Goes," WSGRA, *Proceedings*, 1933; and "President's Address," WSGRA, *Proceedings*, 1934.
35. *Forty Years,* 15.
36. "President's Address," n. 34.
37. Federal spending is the subject of W. H. Lynch paper, *Proceedings,* 1931. County versus state spending is the topic of "Dill Address," *Proceedings*, 1931; "Lynch-Morris Address," *Proceedings*, 1933; and "Highway Director's Report," *Proceedings*, 1934.
38. "Where the Highway Money Goes," and "President's Address," n. 34.
39. NEIC minutes, box 4, file 7, 1 (founding); 12–13 (report); 14 (first quotation); 15 (second quotation). For road problems and private work, see 19, 21, 24. For "all work" quotation, see 67, NEIC, WSGRA Records, CPNS.
40. Concerns of the NEIC may be followed in the NEIC minutes, 79, 93; and in the commissioner's speech, 202 (quotation), box 4, file 7, NEIC, WSGRA Records, CPNS.
41. NEIC minutes, for lack of funding, 93; and for commissioner's speech, 102, NEIC, WSGRA, CPNS.
42. NEIC minutes, 75, NEIC, WSGRA, CPNS.
43. "Balfour Discussion," WSGRA, *Proceedings,* 1–2 October, 1936, Ellensburg.
44. Ibid. For a concrete example in Yakima County, see F. O. Hagie to A. C. Clausen, June 25, 1930, copy in box 7, file 31, ACP, UW.
45. "Horrocks' Presentation," WSGRA, *Proceedings,* 1936.
46. "President's Address," WSGRA, *Proceedings*, 1934.

47. Ibid., and *Forty Years*, 12.
48. Hastings address, WSGRA *Proceedings*, 1932; and "Where the Highway Money Goes," and "President's Address," n. 34.
49. Curtis to Frank W. Guilbert, September 26, 1928, box 5, file 22 (quotation). ACP, UW; and WSGRA, *Proceedings*, 1930. Several letters in ACP, UW discuss Curtis's idea.
50. Curtis to Wellington Pegg, November 15, 1930, box 8, file 10, ACP, UW. Curtis later abandoned his bond scheme. In 1932, in the depths of the Great Depression, he decided that "our safest plan is to stay on a 'pay as you go' basis," Curtis to Frank Guilbert, August 28, 1932, box 4, file 34, ACP, UW.
51. Curtis to Mrs. R. D. Merrill, September 14, 1931, box 9, file 2, ACP, UW.
52. Curtis to T. B. Jones, November 6, 1931, box 9, file 6, ACP, UW.
53. Curtis to Frank W. Guilbert, January 4, 1936, box 11, file 18, ACP, UW.
54. "Where the Highway Money Goes," WSGRA, *Proceedings*, 1933.
55. WSGRA, *Proceedings*, 1932.
56. Curtis to Mark Reed, October 26, 1931, box 9, file 4, ACP, UW.
57. Curtis to Samuel J. Humes, September 3, 1931, box 4, file 1, ACP, UW.
58. Curtis to Frank W. Guilbert, August 22, 1932, box 9, file 34, ACP, UW.
59. Curtis to Senator Wesley L. Jones, February 9, 1932, box 9, file 14, ACP, UW.
60. Curtis to Henry T. Rainey, January 3, 1933, box 10, file 4, ACP, UW.
61. "Where the Highway Money Goes," WSGRA, *Proceedings*, 1933.
62. For Curtis's language, ibid. For Martin's address and report of committee appointed to meet with Martin, see WSGRA, *Proceedings*, 1934.
63. *State ex. rel. Draham v. Cliff Yelle*, 175 Wash. 33; "President's Address," WSGRA, *Proceedings*, 1934; and "Where the Highway Money Goes," WSGRA, *Proceedings*, 1933.

Notes to Chapter Five

1. These issues are discussed in Samuel J. Humes to Curtis, February 23, 1932, box 9, file 12, ACP, UW.
2. For Curtis's concern for trees, see Cox, *The Park Builders: A History of State Parks in the Pacific Northwest* (Seattle: University of Washington Press, 1988), 57–72. Curtis, "Protection of the Highways," typescript, August 27, 1918, box 1, file 6, ACP, UW; and for tree saving agitation, see Herbert Evison to Curtis, February 10, 1921, box 1, file 35, ACP, UW.
3. Curtis to Henry Ball, May 8, 1924, box 2, file 22, ACP, UW.
4. "Report of Organization Meeting, March 3, 1930," box 1, file 8, Eula Lee Merrill Papers, accession 4424-1 (EMP), UW. Original members are listed in box 1, file 1; and attendance and meeting reports in box 1, file 8, EMP, UW.
5. For letter writing, see C. L. Duh to Avis L. Marble, December 30, 1931, box 1, file 16; and L. H. Newell to Merrill, October 22, 1929, box 2, file 6, both in EMP, UW. For other activities, see report, box 1, file 10; box 2, file 2; and Mrs. W. L. Lawton to Merrill, February 25, 1930, box 1, file 6, all in EMP, UW.
6. Curtis to Merrill, August 20, 1931, box 8, file 38, ACP, UW.
7. Curtis to Merrill, August 21, 1931, box 8, file 38, ACP, UW.
8. Curtis to Merrill, September 14, 1931, box 9, file 2, ACP, UW.
9. Curtis to Merrill, February 9, 1932, box 1, file 3, EMP, UW.
10. L. V. Murrow to D. Shelor, October 19, 1933, box 52, file Good Roads Associations and Meetings, Department of Highways Subject File (DHSF) 0.01-0.108, WSA.
11. Typescript 2, box 1, file 1, EMP, UW.
12. *Forty Years*, 23.

13. L. V. Murrow to Irving Clark, November 14, 1938, box 77, file Maintenance–General: Landscaping & Beautification, Drinking Water, DHSF, .0566-11.00, WSA; and Curtis to Ross K. Tiffany, February 16, 1934, box 10, file 24, ACP, UW.
14. Curtis to George F. Yantis, July 14, 1938, box 11, file 24, ACP, UW.
15. Curtis to Tiffany, n. 13.
16. Early truck damage is noted in Goddard, *Getting There*, 95.
17. Alice A. Ridge and John Wm. Ridge, *Introducing the Yellowstone Trail: A Good Road from Plymouth Rock to Puget Sound, 1912–1930* (Altoona, WI: Yellowstone Trail Publishers, 2000), 43–44. Rail and road issues are in Goddard, *Getting There*, 85–119, 138–50.
18. Clay McShane, *Down the Asphalt Path: The Automobile and the American City* (New York: Columbia University Press, 1994), 216. Allen is quoted indirectly in C. C. Finn to Curtis, July 2, 1925, box 2, file 37, ACP, UW.
19. WSGRA, *Proceedings*, September 29–30 and October 1, 1938, Seattle.
20. For the highway patrol, see Curtis to Frank Guilbert, August 26, 1926, box 3, file 31; and for photographs, see Curtis to Mark Reed, October 26, 1931, box 9, file 4, both in ACP, UW.
21. For highway improvements, Curtis to G. E. Hawthorn, July 25, 1934, box 10, file 32; taxes, Curtis to J. A. Snoddy, September 28, 1930, box 8, file 12; reduced loads, Curtis to Nathan Eckstein, February 19, 1927, box 4, file 5; and limiting trucks to certain roads, Curtis to Reed, n. 20, all in ACP, UW.
22. "President's Address," WSGRA, *Proceedings*, 1934. For a comment on railroads, see "President's Address," WSGRA, *Proceedings*, 1933.
23. WSGRA, "President's Address," *Proceedings*, 1934.
24. Goddard, *Getting There*, 117–18, quotation, 118.
25. WSGRA, *Proceedings*, 1938.
26. For background on Parmley, see Ridge and Ridge, *Introducing the Yellowstone Trail*, 1–2; and Harold A. Meeks, *On the Road to Yellowstone* (Missoula, MT: Pictorial Histories Publishing, 2000), 41–42.
27. Ridge and Ridge, *Introducing the Yellowstone Trail*, 10; and Meeks, *Road to Yellowstone*, 42.
28. For the formation and early years of the Trail, see Ridge and Ridge, *Introducing the Yellowstone Trail*, 33–62; and Meeks, *Road to Yellowstone*, 41–62, 69–122.
29. Curtis to B. R. Mandelb, November 28, 1924, box 2, file 27, ACP, UW. For Curtis's election to the executive committee, see *Times*, October 17, 1924.
30. Drake Hokanson, *The Lincoln Highway: Main Street across America*, 10th anniv. ed. (Iowa City: University of Iowa Press, 1999), 5–8.
31. Ibid., 8–13; and Meeks, *Road to Yellowstone*, 165–66.
32. Hokanson, *Lincoln Highway*, 14–17.
33. Ibid., 19, 82–83; and statement from H. O. Cooley to the Seattle Chamber of Commerce accompanying letter, H. O. Cooley to the president, Seattle Chamber of Commerce, February 21, 1927, box 4, file 5, ACP, UW.
34. Curtis to H. L. Weister, November 27, 1925, box 4, file 5, ACP, UW.
35. Ridge and Ridge, *Introducing the Yellowstone Trail*, 16 (first and second quotations); and Meeks, *Road to Yellowstone*, 175 (third quotation).
36. Curtis to J. W. Parmley, May 7, 1927, box 4, file 15.
37. Meeks, *Road to Yellowstone*, 132.
38. Ibid., 132–35, and map, 135, 145 (quotations), 146; and 152. For per-day mileage, see the description of the E. W. Pollock and family trip over roads from Cleveland to Seattle, *Times*, September 28, 1919.
39. Ibid., 144–45, quotation on 145.

40. Curtis to H. L. Weister, n. 34 (first quotation); "Statement to the Finance Committee of the Seattle Chamber of Commerce by the Yellowstone Trail Association, Inc.," box 4, file 5 (second quotation); and Curtis address, WCCO, February 5, 1925, box 1, file 5, all in ACP, UW.
41. Curtis to P. H. Sceva, January 23, 1930, box 7, file 5, ACP, UW.
42. For the 25 percent diversion, see Meeks, *Road to Yellowstone*, 146. Curtis's percentage dated from 1927, after the change in route, so he had nothing to gain by it. See Curtis to Christy Thomas, March 20, 1927, box 4, file 10; and for the "bad condition" quotation, see Curtis to Frank Poole, October 2, 1926, box 3, file 35, both in ACP, UW.
43. For general information on the route change, see Meeks, *Road to Yellowstone*, 55, 146–47, and Yvonne Prater, "The Old Yellowstone Trail," *Columbia* 10 (Spring 1996): 41–42. For the funding, see *P-I*, March 22, 1927.
44. For Curtis's proposal, see Curtis to T. B. Jones, September 1, 1929, box 6, file 22, ACP, UW. The Loop Highway Association caused Curtis many headaches. See Curtis to Paul Sceva, April 24, 1930, box 7, file 21; Curtis to W. H. Little, May 27, 1930, box 7, file 27; and Curtis to Ralph W. Hanson, April 20, 1940, box 12, file 9, all in ACP, UW.
45. "Statement," n. 40; and Curtis to W. L. Weister, October 9, 1925, box 3, file 2, ACP, UW.
46. Christy Thomas to H. O. Colley [sic], February 25, 1927; and Cooley to Thomas, February 28, 1927, copies, both in box 4, file 5, ACP, UW.
47. Curtis to W. L. Weister, n. 45; and Curtis to H. O. Cooley, November 19, 1925, box 3, file 2 (quotations). For efforts to find a replacement, see John A. Gellatley to Curtis, October 13, 1925, box 3, file 2; and Curtis to T. B. Jones [November 5, 1926], box 3, file 37, both in ACP, UW.
48. "Statement," n. 40; and H. O. Cooley to Christy Thomas, copy, October 25, 1926, box 3, file 36, ACP, UW.
49. Curtis to H. O. Cooley, December 23, 1925, box 3, file 27 (quotations); Curtis to Cooley, April 1, 1926, box 3, file 21; and, for trip results, Curtis to T. B. Jones, n. 47, all in ACP, UW.
50. Curtis to J. W. Parmley, n. 36.
51. H. O. Cooley to Curtis, May 9, 1927, box 4, file 15, ACP, UW. For more on the Parmley issue, see Ridge and Ridge, *Introducing the Yellowstone Trail*, 65–68; and Meeks, *Road to Yellowstone*, 174–75.
52. Curtis to J. W. Parmley, n. 36.
53. Hokanson, *Lincoln Highway*, 109–11.
54. Seely, *Building American Highway System*, 78–79; Hokanson, *Lincoln Highway*, 106, 108–09; and Meeks, *Road to Yellowstone*, 171–73.
55. Ridge and Ridge, *Introducing the Yellowstone Trail*, 48–51.
56. Cooley to Curtis, n. 51.
57. Curtis to Platt Morrow, April 7, 1928; and Morrow to Curtis, April 13, 1928, both in box 5, file 11, ACP, UW.
58. Cooley to Curtis, May 31, 1929, box 6, file 13; and Curtis to Cooley, June 5, 1929, box 6, file 14, both in ACP, UW.
59. T. B. Jones to Curtis, October 29, 1929, box 6, file 27, ACP, UW.
60. For Cooley's speech, see Ridge and Ridge, *Introducing the Yellowstone Trail*, 66–67. For Cooley's plans, see Cooley to Curtis, December 18, 1929, box 6, file 32, ACP, UW.
61. The 1927 debt is in Cooley to Curtis, n. 51. For Cooley's letters, see Meeks, *Road to Yellowstone*, 175; and Cooley to Curtis, March 14, 1930, box 7, file 16. Parmley is quoted in Meeks, *Road to Yellowstone*, 175. See also 176.
62. Ridge and Ridge, *Introducing the Yellowstone Trail*, 69–73; and Meeks, *Road to Yellowstone*, 175–77, 179.

63. The 96,000 highway miles are in Meeks, *Road to Yellowstone*, 171.
64. Curtis to H. O. Cooley, April 28, 1926, box 3, file 21, ACP, UW.
65. Curtis to T. B. Jones, October 30, 1929, box 6, file 27, ACP, UW.
66. Department of State, Conference Series No. 14, *Report of the Commission to Study the Proposed Highway to Alaska, 1933* (Washington, DC: 1933), 6–8, quotation on 8; Claus-M. Naske, *Paving Alaska's Trails: The Work of the Alaska Road Commission* (Lanham, MD: University Press of America, 1986), 212–14; and WSGRA, *Proceedings*, 1929. Initial enthusiasm for the road is in *Times*, December 29, 1929.
67. *Report of the Commission,* 15, 29, 43–44, quotation on 44; and the author's observations based on four years' residence in Alaska.
68. *Report of the Commission*, 3, 9, 17, 29, 43, 44.
69. Ibid., 11–19.
70. Ibid., 1, 21, 22 (quotation).
71. WSGRA, *Proceedings*, 1934; and mss. attributed to Carl E. Webber, box 1, file 7, ACP, UW.
72. Malcolm Elliott to Curtis, January 11, 1930, box 7, file 1; Curtis to J. G. Rivers, October 28, 1930, box 8, file 9; and Curtis to Franklin Fisher, September 5, 1930, box 8, file 5, all in ACP, UW.
73. Ernest Walker Sawyer to Curtis, November 7, 1930, box 8, file 4, ACP, UW. Reports on the trip are in *Report of the Commission*, 7; and *Times*, July 13, 1930. For the "very splendid" quotation, see Walter A. Averill to Curtis, March 25, 1931, box 8, file 24, ACP, UW.
74. Ernest Walker Sawyer to Curtis, April 11, 1932, box 9, file 20; J. Gordon Smith to Curtis, April 2, 1932, box 9, file 20; Curtis to Samuel Humes, May 2, 1932, box 9, file 21; and Curtis to Sawyer, May 4, 1932, box 9, file 21, all in ACP, UW.
75. For the later commission, see Webber, n. 71. Curtis to Hill, November 16, 1929, box 6, file 28; ACP, UW; and Sawyer to Hartley, February 20, 1932, box 2K-1-26, HSF, WSA.
76. For the attitude of the Canadian federal government, *Report of the Commission*, 5; and for the Tolmie quotation, 7. Tolmie to Curtis, November 17, 1932, box 2L-1-43, Subject Files (SF), Clarence D. Martin Papers (CMP), WSA.
77. J. Gordon Smith to Curtis, November 13, 1930, box 8, file 10, ACP, UW; and Tolmie to Curtis, n. 76.
78. *Report of the Commission*, 23–31; 35–38.
79. For "settled portion" quotation, see Curtis to George Ogsten, March 3, 1934, box 10, file 24. For "the province" quotation, see S. F. Tolmie to Curtis, September 29, 1934, box 10, file 36. For "very greatly" quotation, see Curtis to Mrs. D. Goodrich, March 20, 1934, box 10, file 25. For the remaining quotations, see Curtis to Ernest Walker Sawyer, July 3, 1935, box 11, file 11, All are in ACP, UW.
80. Heath Twichell, *Northwest Epic: The Building of the Alaska Highway* (New York: St. Martin's, 1992), 18–20.
81. Curtis to Frank W. Guilbert, July 11, 1940, box 12, file 11, ACP, UW.
82. K. S. Coates and W. R. Morrison, *The Alaska Highway in World War II: The U.S. Army of Occupation in Canada's Northwest* (Norman: University of Oklahoma Press, 1992), 18–37.
83. Orlando W. Miller, *The Frontier in Alaska and the Matanuska Colony* (New Haven: Yale University Press, 1975), especially 1–13.

Notes to Chapter Six

1. Theodore Catton, *National Park, City Playground: Mount Rainier in the Twentieth Century* (Seattle: University of Washington Press, 2006), 10–12, 33; and Aubrey L. Haines, *Mountain Fever: Historic Conquests of Rainier,* reprint ed. with forward by Ruth Kirk (1982; reprint, Seattle: University of Washington Press, 1999), 14, 29–30, 35–40, 158–62, 168–70.

2. Haines, *Mountain Fever*, does not mention Curtis among the 1896 climbers, 189–94.
3. "President's Address," WSGRA, *Proceedings*, 1934. Superintendent's Annual Report (SAR), 1933; and SAR, 1935, box 2, Record Group 79, Records of Mount Rainier National Park (RG79), Pacific Alaska Region, Seattle (PARS), National Archives and Records Administration (NARA).
4. Catton, *National Park, City Playground*, 12; and Haines, *Mountain Fever*, 153–55.
5. Roderick Frazier Nash, *Wilderness and the American Mind*, 4th ed. (New Haven: Yale University Press, 2001), 1–160.
6. Nash, *Wilderness and American Mind*, 156.
7. Ibid., 141–60.
8. Catton, *National Park, City Playground*, 12–13, 15–25. An excellent popular history of the relationship between railroads and national parks is Alfred Runte, *Trains of Discovery: Railroads and the Legacy of Our National Parks*, 5th ed. (Lanham, MD: Roberts Rinehart, 2011).
9. Catton, *National Park, City Playground*, 17–18; and Nash, *Wilderness and American Mind*, 113–21. See Catton, 25–31, for the park bill's passage through Congress.
10. Catton, *National Park, City Playground*, 32–37; 42–45, 47. See also Bette Filley, *The Big Fact Book About Mount Rainier* (Issaquah, WA: Dunamis House, 1996), 56; and reproduction of the original park act, 401–05; Ruth Kirk, *Sunrise to Paradise: The Story of Mount Rainier* (Seattle: University of Washington Press, 1999), 59; and David Louter, *Windshield Wilderness: Cars, Roads, and Nature in Washington's National Parks* (Seattle: University of Washington Press, 2006), 14.
11. Robert Shankland, *Steve Mather of the National Parks*, 3rd ed. rev. and enl. (New York: Knopf, 1970), 77.
12. Haines, *Mountain Fever*, 24, 26, 43, 53, 54–55, 65, 67, 92, 96, 100, 113, 136, 165, and for quotation, 191–92.
13. Bruce Barcott, *The Measure of a Mountain: Beauty and Terror on Mount Rainier* (Seattle: Sasquatch, 1997), 156–69.
14. Haines, *Mountain Fever*, 196–97.
15. Minutes of the meeting of November 26, 1908, microfilm reel 1, Minute Books, Mountaineers Club of Seattle, accession 1984–3, UW (hereafter cited by date, reel, and Mountaineers); and William Frederick Bade, "On the Trail with the Sierra Club," *Sierra Club Bulletin* 5 (January 1, 1904): 50–65.
16. Shankland, *Steve Mather*, 9.
17. William A. Brooks, "With Sierrans and Mazamas—July, 1905," *Appalachia* 11 (May 1906): 114–25; and Evelyn Marianne Ratcliff, "The Sierra Club's Ascent of Mt. Rainier," *Sierra Club Bulletin* 6 (January 1906): 1–5, and plate 7 following p. 5. See also "Report of the Outing Committee" in the same *Sierra Club Bulletin*, 50–51. For Price, see the *P-I* feature, Frank Lynch, "The Conquest of Our Own Mount Everest," March 23, 1955, undated copy in box 3, file Curtis Mountains, AS, APRCA, UAF, and below. For Mather, see Shankland, *Steve Mather*, 3–113, especially 9, 47, 61–62, and below.
18. Brooks, "With the Sierrans and Manzamas," 124, 125 (quotations).
19. Shankland, *Steve Mather*, 37 (quotation), 107–12, 221–24; and Frederick K. Goodwin and Kay Redfield Jamison, *Manic-Depressive Illness: Bipolar Disorders and Recurrent Depression*, 2nd ed. (New York: Oxford University Press, 2007), xix–xxiii, 3–27, 29–87, 798.
20. Lynch, "Conquest of Our Own Everest," n. 17.
21. Curtis, "The First Ascent of Mount Shuksan," *Mountaineer* 1 (June 1907): 52. The photograph is in Frederick and Engerman, *Asahel Curtis*, outside back cover.
22. The foundation lore is in Jim Kjeldsen, *The Mountaineers: A History* (Seattle: Mountaineers, 1998), 11. The "small club...twenty folks" quotation is from Mollie Leckenby

King, "Early Outings Through the Eyes of a Girl," *Mountaineer* 50 (December 1956): 36. For the quotations from the provisional club, see Minutes, November 8, 1906, reel 1, Mountaineers, UW.

23. *P-I*, November 10, 1906.
24. Robert M. Bryce, *Cook & Peary: The Polar Controversy Resolved* (Mechanicsburg, PA: Stackpole, 1997), 235–92, 795–844; Bruce Henderson, *True North: Peary, Cook, and the Race to the Pole* (New York: Norton, 2008), 145–55, 173–79, 281–84; and William R. Hunt, *To Stand at the Pole: The Dr. Cook–Admiral Peary North Pole Controversy* (New York: Stein and Day, 1981), 21–32, 105–11, 139–91, plus "peak" photos between 142–45 and 213–21.
25. Minutes, November 16, 1906, December 18, 1906, January 18, 1907, and November 15, 1907, reel 1, Mountaineers, UW. For the end of 1907 membership, see Kjeldsen, *Mountaineers*, 14.
26. Kjeldsen, *Mountaineers*, 14 (quotation). See also 12–14, 16–19, 22, and photo, 13.
27. Sierra Club, www.sierraclub.org/history/origins; and Mazamas, www.mazamas.org. For Mountaineers' requirements, see Kjeldsen, *Mountaineers*, 14, 19. The requirements later were dropped in favor of admitting anyone over fourteen years of age who was willing to pay the initiation fee and dues.
28. Curtis's relations with the Forest Service are in Special Use Agreement (Permit), June 27, 1908; Curtis to A. J. Craven, June 22, 1908, both in box 1, file 2, ACP, UW; Curtis to G. F. Allen, August 17, 1908; and Curtis to G. F. Allen, September 25, 1908, both in box 20, RG 79. Park Central Files (old), PARS, NARA. For the quotation, see H. L. Hurd to T. E. O'Farrell, June 18, 1909, box 20, RG 79, PARS, NARA.
29. Kjeldsen, *Mountaineers*, 22–26.
30. Curtis, "Mountaineers' Outing to Mount Rainier," *Mountaineer* 2 (November 1909): 5–6. See also "Climbing Rainier with Curtis in 1909," *Columbia* 1 (Summer 1987): 29–37 [includes Curtis photographs].
31. The Curtis Papers contain numerous letters concerning the organization of the outing. For examples see Curtis to Henry Loss, January 6, 1909, box 1, file 7; W. A. Ross to Curtis, December 8, 1908, box 1, file 7; contract between Curtis and Ross, January 26, 1909, box 1, file 7; and Curtis to F. M. Fultz, April 16, 1909, box 1, file 8, all in ACP, UW. For the average income, see *Historical Statistics of the United States: Colonial Times to 1970* (Washington, DC: Bureau of the Census, 1975), 1: 172.
32. Curtis, "Mountaineers' Outing," 6 (first two quotations), 7 (all other quotations). Curtis's broken shoulder blade is recalled in L. A. Nelson, "Thirty Years in Retrospect: The First Decade in Mountaineer Annals," *Mountaineer* 30 (December 1937): 11.
33. The organization of the climb is in Curtis, "Mountaineers' Outing," 9–10. Curtis's recollection is in Curtis to Miss Lulie Nettleton, March 20, 1936, Box 11, file 19, ACP, UW.
34. Curtis, "Mountaineers' Outing," 8 (first two quotations), 9 (all other quotations). For Curtis and the flag, see James A. Wood to Curtis, July 16, 1909, box 1, file 12, ACP, UW.
35. Curtis, "Mountaineers' Outing," 9–11. For Curtis's reaction to Spray Park, see 11 ("so beautiful" quotation); and 12 ("trails" and "strong hand" quotations).
36. Ibid., 4.
37. Curtis to Meany, July 24, 1909, box 24, file 21, ESMP, UW. For Curtis's reelection, see Minutes, November 18, 1909, reel 1, Mountaineers, UW. Meany to Curtis, July 17, 1910, box 1, file 8, ACP, UW.
38. Lulie Nettleton, "Mountaineers' Outing on Glacier Peak," *Mountaineer* 3 (November 1910): 29–40; and Winona Bailey, "The Mt. Adams Outing of 1911," *Mountaineer* 4 (1911): 24. The accounting controversy is in *Times*, July 6, 1910.

39. For analyses of Curtis's relation to the Mountaineers, hostile to Curtis, see Kjeldsen, *Mountaineers*, 120, 122; and Carsten Lien, *Olympic Battleground: The Power Politics of Timber Preservation*, 2d ed. (Seattle: Mountaineers, 2000), 37–44.
40. Don Duncan, "Curtis, Asahel, 1874–1941, Photographer," essay 8780, History Link, www.historylink.org.
41. Eaton to Meany, April 5, 1911, copy, box 1, file 16, ACP, UW.
42. Meany to Eaton, April 11, 1911, letter press book 8, ESMP, UW.
43. Meany to Curtis, September 1, 1911; and Curtis to Meany, September 3, 1911, both in box 1, file 17, ACP, UW.
44. Curtis to Fisher, September 7, 1911; Mount Rainier Scrapbooks (MRS), vol. 1; microfilm; accession 4958-002 (ACP2); Fisher to Curtis, November 9, 1911, box 1, file 17, ACP; and Curtis to J. C. Slater, November 18, 1911, MRS, vol. 1, ACP2; all in UW. *P-I*, November 18, 1911. Curtis to Meany, November 17, 1911, box 1, file 17, ACP, UW.
45. George A. Frykman, *Seattle's Historian and Promoter: The Life of Edmond Stephen Meany* (Pullman: Washington State University Press, 1998), 198.
46. Vol. 4, pp. 42–43. All quotations except the one beginning "the true" are on 42. The final quotations are on 43.
47. Frykman, *Seattle's Historian*, 1–142; and photo section between 132 and 133.
48. Chas. H. Flory to F. H. Stannard, May 11, 1909; and R. A. Ballinger to F. H. Stannard, May 14, 1909, copies, both in box 1, file 8, ACP, UW. See also excerpt and copy of letter to Curtis, possibly 1915, from a Forest Service representative, box 1, file 28, ACP, UW.
49. Betty McCullough to Archie Satterfield, September 20, 1969, box 3, file Curtis Mountains, AS, APRC, UAF; and Nash, *Wilderness and American Mind*, 44–141.
50. "Suggested plan for the concerted action in the matter of the Olympic Natl Monument," n.d., box 1, file 17, ACP, UW; and [F. H. Stannard] to Wm. Sulzer, February 23, 1912, box 1, file 19, ACP, UW.
51. For the monument's creation and related issues, see Robert L. Wood, *The Land that Slept Late: The Olympic Mountains in Legend and History* (Seattle: Mountaineers, 1995), 93–95, 131–35, 156–59; Gail E. H. Evans, *Historic Resource Study, Olympic National Park, Washington* (Seattle: National Park Service, 1983), 1–57, 149; Tim McNulty, *Olympic National Park: A Natural History* (Seattle: University of Washington Press, 2003), 207–13; and Ben W. Twight, *Organizational Values and Political Power: The Forest Service Versus the Olympic National Park* (University Park: Pennsylvania State University Press, 1983), 33–38. Curtis to E. T. Parsons, May 21, 1912, box 1, file 22 ("mining interests" quotation); to Enos A. Mills, September 1, 1912, box 1, file 22 ("to have" quotation); and to Quilcene Improvement Club, January 31, 1912, box 1, file 18 ("quite a number" and subsequent quotations), all in ACP, UW.
52. Wood, *The Land that Slept Late*, 156–57, and other works cited in n. 51.
53. Curtis to Grant W. Humes, March 4, 1909, box 1, file 7, ACP, UW.
54. Curtis to Quilcene Improvement Club; and to E. T. Parsons, n. 51.
55. Curtis to Mills (first quotation); and to Parsons ("I went" and second "commercial bodies" quotations), n. 51. Quotations from the report are in *Times*, January 14, 1912.
56. Edmond S. Meany, "The Olympic National Monument," *Mountaineer* 4, 58–59.
57. E. T. Parsons to Curtis, April 8, 1912, box 1, file 21, ACP, UW.
58. Richard West Sellers, *Preserving Nature in the National Parks: A History* (New Haven: Yale University Press, 1977), 36, 65.
59. Curtis to Nelson, February 7, 1912, box 1, file 18, ACP, UW.
60. Nelson to Curtis, February 11, 1912, box 1, file 18, ACP, UW.

61. Curtis to Nelson, n. 59 (first quotation); and to Mills, n. 51 (second quotation). For Curtis's continued efforts, see Curtis to Will D. Pratt, November 4, 1915; box 1, file 28; and Twight, *Organizational Values and Political Power*, 37–38.
62. Curtis to Nettleton, n. 33 ("tightly bound" quotation); and King, "Early Outings," 36 ("Many Mountaineers" quotation).
63. Curtis to Nelson, n. 59; Curtis to Nelson, February 14, 1912, box 1, file 18; and Curtis to Wm. E. Colby, May 18, 1912, box 1, file 20, both in ACP, UW.
64. [F. H. Stannard] to Wm. Sulzer, n. 50.
65. Kjeldsen, *Mountaineers*, 120 (first quotation); and Lien, *Olympic Battleground*, 44 (other quotations).

Notes to Chapter Seven

1. *Seventeenth Decennial Census of the United States, Census of Population,* 1950, vol. 2, pt. 47 (Washington, DC: 1950), table 4, p. 47–48.
2. Curtis to Walter L. Fisher, September 7, 1911, vol. 1, microfilm, ACP2; and Curtis to E. S. Meany, September 3, 1911 (quotations), box 1, file 17, ACP, both in UW.
3. *P-I*, January 18, 1912; and for the response, see T. H. Martin to Curtis, January 18, 1912, and other letters in box 13, MRS, vol. 1, ACP, UW.
4. For developments, see R. A. Ballinger to Board of Trustees, New Seattle Chamber of Commerce, March 7, 1912; and Memorandum from Joint Committee, March 7, 1912 (quotations), both in box 13, MRS, vol. 1, ACP, UW. For the final organization see clipping attributed to the *Tacoma Ledger* (*Ledger*) March 15, 1912, vol. 1, microfilm, ACP2, UW. For Ricksecker and Mather, see Louter, *Windshield Wilderness*, 13–19, 39–49.
5. T. H. Martin to *Ledger*, April 3, 1912; Martin to Tacoma Members, April 4, 1912; Curtis to E. S. Hall, April 16, 1912; and Martin to Curtis, April 25, 1912, all in MRS, vol. 1, microfilm, ACP2, UW. Letters in box 13, vol. 1, ACP, UW confirm these developments.
6. For the name change, see Curtis to General James A. Drain, August 18, 1913, vol. 4, microfilm, ACP2, UW. For later organizational developments, Curtis and Martin to Stephen T. Mather, January 23, 1919; *Times*, March 19, 1920; and clipping from the *Chehalis Bee-Nugget*, August 20, 1920, both in box 15, MRS, vol. 7, ACP, UW. See also Curtis to Mather, January 12, 1921, box 1, file 34, ACP, UW.
7. Curtis to W. H. Peters, November 20, 1920 (quotation), box 15, MRS, vol. 7, ACP, UW; and Catton, *National Park, City Playground*, 61–81 For long-range goals, see Curtis to Harry Weir, November 15, 1920; box 15, vol. 7 (appropriations); and Martin to Lane, July 26, 1913, box 14, MRS, vol. 4 (advice and quotation), both in ACP, UW. For their part, the park service and the interior department allowed Curtis to announce road improvements to Rainier, *Times*, November 7, 1927, and August 4, 1932.
8. For an example of the unified view, see Curtis to A. J. Richie, February 10, 1927, box 4, file 4, ACP, UW. An example of a Curtis initiative, in this instance grazing in Rainier, is in W. H. Peters to Curtis, January 4, 1921; Curtis to Peters, January 8, 1921; and Curtis to Harry L. Myers, January 22, 1921, box 1, file 34, ACP, UW. For a board meeting, see minutes, n.d. but June 10, 1927, box 5, file 15, ACP, UW.
9. Catton, *National Park, City Playground*, 50, 62–69.
10. Ibid., 67–81; Lien, *Olympic Battleground*, 72–77; and G. F. Allen, *Report of the Acting Superintendent of the Mount Rainier National Park to the Secretary of the Interior, 1908* (Washington, DC: 1908), 15.
11. SAR, 1924, pp. 1, 2 (quotation), box 2, RG 79, PARS, NARA.
12. Catton, *National Park, City Playground*, 78, 80–81, 90–91; and Martin to Orpheus C. Soots, December 8, 1924, copy, box 2, file 28, ACP, UW.

13. Curtis to Tacoma [and] Eastern R. R. Co, September 16, 1909 ("possible" quotation), box 1, file 14; and Curtis to Edward S. Hall, February 18, 1911 ("would" quotation), box 1, file 16, both in ACP, UW.
14. Curtis to Mather, December 27, 1917; and Curtis to Albright, September 24, 1917, both in box 1, file 30, ACP, UW.
15. Curtis to Mather, December 7, 1917; Curtis to Mather, December 27, 1917; and Mather to Curtis, December 15, 1917, all in box 1, file 30, ACP, UW. For praise of the ranger, Curtis to Horace M. Albright, September 24, 1917, box l, file 30, ACP, UW.
16. Albright to Curtis, August 24, 1918, box 1, file 30; Mather to Curtis, October 22, 1918, box 1, file 31; and Curtis to Mather, August 10, 1918, box 1, file 31, all in ACP, UW.
17. Curtis to Mather, August 10, 1918, box 1, file 31, ACP, UW.
18. An example of Martin's hard work is Martin to Curtis, October 7, 1912, vol. 1, microfilm, ACP2, UW.
19. Martin to Board of County Commissioners of Pierce County, April 18, 1913, vol. 4, microfilm, ACP2, UW.
20. Graves's attitude is in Lancaster to Martin, January 10, 1913, copy, vol. 1, microfilm, ACP2, UW. For Lancaster's work, see Lancaster to Curtis, January 11, 1913, box 13, MRS, vol. 3, ACP, UW; clipping, *Ledger*, January 25, 1913; unattributed clipping, (quotation) April 12, 1913, vol. 4, microfilm, ACP2, UW; and Curtis to Major J. B. Cavanaugh, April 24, 1913, box 20, file 152, RG 79, PARS, NARA.
21. The years 1921–1924 saw a large correspondence. Examples are Curtis to Mather, February 21, 1921, box 1, file 35; Curtis to George W. Goodwin, June 14, 1923, box 2, file 12 (renege); Curtis to Horace Whitacre, April 17, 1924, box 2, file 19 (Curtis's belief); Curtis to C. L. Nelson, July 28, 1923, box 2, file 13 (quotation); and Curtis to O. A. Tomlinson, May 8, 1924, box 2, file 22, all in ACP, UW.
22. Curtis to Romans, January 31, 1914, box 1, file 27, ACP, UW.
23. A comprehensive account of the issue, although biased in favor of Tacoma, is Genevieve McCoy, "'Call It Mount Tacoma': A History of the Controversy Over the Name of Mount Rainier," (master's thesis, University of Washington, 1984).
24. Curtis to Senator [sic] Johnson, February 27, 1924, box 2, file 17; Curtis to Nicholas Sinnott, April 21, 1924, box 2, file 20; and other letters in box 2, files 17–35, ACP, UW.
25. Curtis to Dill, February 27, 1924, box 2, file 17; Curtis to Jones, April 21, 1924, box 2, file 20; Curtis to Dill, April 22, 1924, box 2, file 20; and Dill to Curtis, April 28, 1924, box 2, file 21, all in ACP, UW. An example of a conciliation is Curtis to George E. Goodman, June 28, 1923, box 2, file 13, ACP, UW.
26. Fred A. Britten to Curtis, January 25, 1925, box 2, file 30; Wm. G. Colby to Curtis, May 10, 1924, box 2, file 27; and Lin H. Hadley to Curtis, box 2, file 21, all in ACP, UW. For Underwood, see *Seattle Town Crier*, June 7, 1924.
27. "Recollections by Hazel Wood Niendorff," box 3, file Curtis Mountains, AS, APRC, UAF.
28. For Ball's action, see Curtis to Whitacre, n. 21. See also Curtis to T. H. Martin, April 18, 1924 (first quotation), box 2, file 19; Curtis to Whitacre, April 19, 1924, box 2, file 19; and Curtis to O. A. Tomlinson, May 8, 1924 (second quotation), box 2, file 22, all in ACP, UW.
29. Whitacre to Curtis, July 11, 1924, box 2, file 23; Curtis to Whitacre, ("explained" quotation), August 25, 1924, box 2, file 24; and Whitacre to Curtis, August 28, 1924, ("appreciating" quotation), box 2, file 24, all in ACP, UW.
30. Curtis to E. G. Griggs, September 27, 1924, box 2, file 25, ACP, UW.
31. The correspondence on these matters is voluminous. See Curtis to T. H. Martin, November 20, 1924; box 2, file 27; Martin to Curtis, December 12, 1924, box 2, file 28; Curtis to

Christy Thomas, March 2, 1925 (first quotation), box 2, file 33; Curtis to Thomas, March 30, 1925, box 2, file 33; and Curtis to Arno B. Cammerer, March 31, 1925, box 2, file 33, all in ACP, UW. For the visit, see Curtis to Arno B. Cammerer, May 28, 1925, box 2, file 35, ACP, UW. For the discussion, see Curtis to Herbert Evison, June 8, 1925, box 2, file 36; and Evison to W. D. Lyness, June 11, 1925, box 2, file 36, both in ACP, UW. For "steady" quotation, see Cammerer to Curtis, June 13, 1925, box 2, file 36, ACP, UW.

32. Accounts of the trip are in Curtis to B. H. Burrell, August 18, 1925; and O. A. Tomlinson to Director, National Park Service, August 24, 1925, copy, both in box 2, file 38, ACP, UW. For river flooding, see Bert H. Burrell to Horace M. Albright, August 15, 1925, copy, box 2, file 38, ACP, UW.
33. T. H. Martin to Curtis, February 19, 1929, box 6, file 3, ACP, UW.
34. For the state system, see H. A. Rhodes to Horace M. Albright, May 13, 1929, copy, box 6, file 12, ACP, UW. On the campaign to extend the highway system, see Curtis to Christy Thomas, September 27, 1926, box 3, file 34; Curtis to Horace M. Albright, February 28, 1929, box 6, file 5; Curtis to T. A. Stevenson, March 9, 1927, box 4, file 8; Curtis to Julius C. Hubbell, February 23, 1929, box 6, file 5; and Curtis to Louis C. Cramton, March 16, 1929, box 6, file 7, all in ACP, UW. For approach roads, see O. A. Tomlinson to The Director, February 25, 1929; and Cramton to Albright, February 28, 1929 (quotations), both in RG 79, box 22, file 152, PARS, NARA.
35. Griggs to T. H. Martin, June 7, 1929 (first quotation), box 22, file 152, RG 79, PARS, NARA. See also O. A. Tomlinson to Director, April 16, 1929; H. A. Rhodes to Horace M. Albright, May 9, 1929; and Martin to Tomlinson, June 13, 1929 (second quotation), all in box 22, file 152, RG 79, PARS, NARA.
36. Cramton to Curtis, April 18, 1928, box 5, file 12; and Underwood to Paul Sceva, November 29, 1929, box 6, file 30, both in ACP, UW.
37. Curtis to Underwood, March 26, 1931, box 8, file 24, ACP, UW; and Catton, *National Park, City Playground*, 61–62, 65–69, 86, 88–89.
38. Catton, *National Park, City Playground*, 90–91, 93–98; and Louter, *Windshield Wilderness*, 62–63. For the litany, see, for example, T. H. Martin to J. J. Underwood, February 18, 1928, copy, box 2, file 32, ACP, UW.
39. Louter, *Windshield Wilderness*, 11–19, 52–53. Curtis to T. H. Martin, January 30, 1925, box 2, file 30; and Curtis to J. A. McKinnon, April 6, 1927, box 4, file 12, both in ACP, UW.
40. These issues are raised in T. H. Martin to Curtis, February 25, 1925, box 2, file 32; and Curtis to Edmond S. Meany, December 20, 1929, box 6, file 34, both in ACP, UW; and Underwood to Sceva, n. 36. For the expense, see O. A. Tomlinson to Curtis, October 14, 1929, box 6, file 25, ACP, UW.
41. Louter, *Windshield Wilderness*, 65–66.
42. Curtis to RNPAB, January 19, 1936, box 10, file 22; and Cammerer to Curtis, March 25, 1936, box 11, file 20, both in ACP, UW.

Notes to Chapter Eight

1. Samuel C. Lancaster, *Report of Samuel C. Lancaster, Commissioner, Seattle-Tacoma Rainier National Park Committee, Services Performed, Sixty-Second Congress, Washington, D.C.* [Seattle, 1913], UW; and clippings in vols. 3 and 4, microfilm, ACP2, UW.
2. For Drain, see T. H. Martin to Senator Wesley L. Jones, July 11, 1914, box 15, MRS, vol. 5, ACP, UW; and Raymond Moley Jr., *The American Legion Story* (New York: Duell, Sloan and Pearce, 1966), 401.
3. The earliest mention of Underwood is Curtis to Stephen T. Mather, January 18, 1921, box 1, file 34, ACP, UW. All of Underwood's letters addressed to others in ACP, UW, are copies. Underwood's reports include letters to Christy Thomas in box 2, November 27, 1923, file

16; April 4, 1924, file 19; and February 12, 1925, file 31, ACP, UW. For Underwood's visit with Coolidge, see Curtis to Horace M. Albright, March 7, 1927, box 4, file 7, ACP, UW.

4. Curtis to Underwood, February 12, 1930, box 7, file 9, ACP, UW.
5. Albright's quotation is in his letter to Curtis, March 2, 1927, box 4, file 7, ACP, UW. For Albright's responses to criticism, see Donald C. Swain, *Wilderness Defender: Horace M. Albright and Conservation* (Chicago: University of Chicago Press, 1970), 136–40. Curtis's quotations are in his response to Albright, March 7, 1927; and to Tomlinson, March 7, 1927, both in box 4, file 7, ACP, UW. For Curtis's restraint, see J. J. Underwood to Curtis, January 19, 1928, box 5, file 3, ACP, UW.
6. O. A. Tomlinson to Curtis, May 14, 1927, box 4, file 10; J. A. Elliot to Curtis, June 16, 1927, box 4, file 16; and Curtis to Thomas H. MacDonald, July 16, 1927, box 4, file 22, and other letters in ACP, UW.
7. For previous requests, see Curtis to T. H. Martin, February 25, 1925, box 2, file 32; for Curtis quotations, Curtis to Stephen T. Mather, January 24, 1926, box 3, file 10; and for Albright quotation, Horace M. Albright to James G. Newbegin, April 30, 1931, copy, box 8, file 29, all in ACP, UW.
8. The percentage is in Curtis to E. F. Benson, n.d. but April 18, 1931, box 8, file 27, ACP, UW. Conflicting routes are in Curtis to RNPAB, April 20, 1931, marked "not sent"; and Curtis to L. A. Nelson, April 23, 1931, both in box 8, file 27, ACP, UW. The park ranger is quoted in James Morton Turner, *The Promise of Wilderness: American Environmental Politics Since 1964* (Seattle: University of Washington Press, 2012), 52.
9. Albright to Tomlinson, May 21, 1931, box 8, file 37; Tomlinson to Curtis, February 19, 1932, box 9, file 16 (quotations) both in ACP, UW. For defeating the relocation, see Curtis to Auburn Chamber of Commerce, September 15, 1935, box 10, file 17; and P. H. Sceva to Curtis, September 19, 1933, both in box 10, file 17, ACP, UW.
10. Curtis to Cammerer, October 16, 1933, box 10, file 19, ACP, UW.
11. For early cooperation with the Mountaineers, see Curtis to E. S. Hall, November 28, 1911, box 1, file 17, ACP, UW; Curtis to E. S. Meany, November 5, 1913, box 28, file 30, ESMP, UW; and John F. Miller to Curtis, January 24, 1921, box 1, file 34, ACP, UW. For 1925, see Curtis to George E. Goodwin, March 30, 1925, box 2, file 33; and meetings of March 5, 1925; and September 3, 1925, box 1, file 6, Minutes of the Board of Trustees, (minutes), Mountaineers, Administrative Division, accession 2885-6, (MAD), UW. For 1927 see J. F. Beede to Curtis, June 21, 1927; and Curtis to Beede, June 22, 1927, both in box 4, file 19, ACP, UW.
12. Curtis to Meany, March 8, 1928; file 7; and Meany to Curtis, March 17, 1928, file 9, both in box 5, ACP, UW.
13. An example of exasperation is in Curtis to Horace M. Albright, December 20, 1926, box 3, file 40, ACP, UW. For keeping the Mountaineers informed, see minutes, November 8, 1928, file 7; and February 6, 1930, file 9, both in box 1, MAD, UW.
14. Cox, *Park Builders*, 73–77; Shankland, *Steve Mather*, 284; and Louter, *Windshield Wilderness*, 284. Robert Sterling Yard to Curtis, September 21, 1929, (quotation), box 6, file 23, ACP, UW.
15. The collaboration is inferred from the fact that the RNPAB and the Forest Service each credited the other with the parkway plan. For the mutual credit, see *Times*, May 12, 1929; and Curtis to J. J. Underwood, May 10, (quotations), 1929, box 6, file 12, ACP, UW. For formal board approval, see Curtis to L. F. Kneipp, May 14, 1930, box 7, file 25, ACP, UW.
16. Hammond to Curtis, September 1, 1929, box 6, file 21; Curtis to David Whitcomb, April 10, 1930 (quotation), box 7, file 19; and, for encouragement, Hammond to Curtis, May 28, 1930, box 7, file 27, all in ACP, UW. Curtis's later assessment of Hammond is in *Times*, June 12, 1936.

17. Shankland, *Steve Mather*, 287–88, 290–91; Planning details are in Curtis to Frank Jones Clark, May 19, 1931, box 8, file 30; Curtis to Hartley, June 17, 1932, box 9, file 25; and Hartley to Curtis, box 9, file 29, all in ACP, UW. For the ceremony, see Curtis to Franklin Adams, July 5, 1932; and to Jane Mather, July 5, 1932, both in box 9, file 30, ACP, UW.
18. Vol. 11, pp. 8, 19, 21. The issue was planned before Mather's death, see D. Shelor to Curtis, January 15, 1930, box 7, file 3, ACP, UW.
19. Curtis to Douglas Shelor, July 26, 1926, box 3, file 25; Curtis to O. A. Tomlinson, June 15, 1927, box 4, file 9; and Curtis to Horace M. Albright, March 15, 1928, box 5, file 8, all in ACP, UW.
20. Sellers, *Preserving Nature*, 80–82.
21. Filley, *Big Fact Book*, 140, 141; and *Times,* September 3, 1933.
22. John W. Summers to Curtis, January 25, 1921, box 1, file 24; and Martin F. Smith to Curtis, June 13, 1933, box 10, file 9, both in ACP, UW.
23. Curtis to Dix H. Roland, September 13, 1921, box 1, file 29, ACP, UW.
24. T. H. Martin to Stanton Warburton, April 29, 1912 (copy), box 13, vol. 1; and Warburton to Curtis, March 17, 1916, box 1, file 29, both in ACP, UW.
25. Curtis to W. D. Lyness, August 31, 1925, box 2, file 39, ACP, UW. *Argus*, December 11, 1926; and December 14, 1929.
26. Curtis, *Rainier National Park*, [Seattle: 1921].
27. *Yakima Herald*, February 14, 1913, box 14, MRS, vol. 3, ACP, UW.
28. Curtis to Horace M. Albright, March 29, 1932, box 9, file 19, ACP, UW.
29. Curtis to Thomas B. Hill, September 22, 1921, box 1, file 39, ACP, UW.
30. Curtis to Harry L. Meyers, January 22, 1921, box 1, file 24, ACP, UW.
31. Curtis to Frank W. Guilbert, November 2, 1920, box 15, file 9, ACP, UW.
32. Curtis to Robert Moran, November 26, 1929, box 6, file 30, ACP, UW. For Curtis's proposal, see *Times,* July 12, 1925, and for his frustrations with developing state parks, see Cox, *Park Builders*, 57–72.
33. Frank W. Guilbert to Curtis, April 18, 1917, box 1, file 27, ACP, UW.
34. Curtis to Robert Sterling Yard, September 25, 1929, box 6, file 23, ACP, UW. For the state chamber, see *Times,* June 22, 1928, March 12, 1929, and December 7, 1938.
35. Stevenson to Curtis, April 15, 1936, box 11, file 20, ACP, UW.
36. Curtis to Stevenson, April 17, 1936; Curtis to Arno B. Cammerer, April 17, 1936; and Curtis to All Members of the [RNPAB], April 18, 1936, all in box 11, file 21, ACP, UW.
37. The extensive correspondence on this issue includes T. A. Stevenson to O. A. Tomlinson, May 28, 1936 (first quotation), box 1, file 4, RG 79, PARS, NARA; Curtis to Stevenson, January 8, 1937; Curtis to H. J. Gille, February 24, 1937; Curtis to Stevenson, February 24, 1937 (second quotation), and attachment (third quotation), all in box 11, file 28, ACP, UW.
38. Copies are in box 1, file 4, accession 4058-3 (ACP3), UW; and box 75, file Olympic Peninsula—reports #5, Records of the State Planning Council (SPCR), WSA.
39. Curtis to Stevenson, n. 36.
40. Stevenson to Curtis, n. 35 (quotation); and Stevenson to Curtis, May 20, 1936, box 1, file 4, RG 79, PARS, NARA.
41. Martin to Mrs. M. G. Mitchell, September 15, 1924 (copy), box 2, file 24, ACP, UW.
42. Martin to Curtis, July 1, 1927, box 4, file 20, ACP, UW.
43. Tomlinson to Curtis, October 1, 1929, box 6, file 24, ACP, UW.
44. W. C. Mumau to Curtis, February 14, 1931, box 8, file 20, ACP, UW.
45. Stevenson to Curtis, January 24, 1933, box 10, file 5, ACP, UW.
46. Albright to Curtis, July 11, 1933, box 10, file 11, ACP, UW.

Notes to Chapter Nine

1. [Curtis], "The Mountaineers' Annual Outing, 1907," *Mountaineer* 1 (March 1907): 45–47; and Curtis, "Storm Bound on Mount Olympus," *Mountaineer* 1 (September 1907): 69–72. Descriptions not cited to other sources are from the second article.
2. Curtis, "Storm Bound," 69.
3. Ibid.
4. Ibid., 71.
5. Ibid.
6. Ibid. See also *Times*, August 15, 1907.
7. Curtis, "Storm Bound," 71–72.
8. *Times*, August 15, 1907; L. A. Nelson, "The Ascent of Mount Olympus," *Mountaineer* 1 (September 1907): 65–68; Mary Banks, "Mountaineers in the Olympics," *Mountaineer* 2 (November 1908): 75–86; and Curtis, "The First Ascent of Mt. Olympus: Climbing the Snow-Clad Olympic Mountains which Overlook the Pacific Ocean," *World's Work* 16 (May 1908): 10,261-62.
9. All quotations are from Curtis, "Storm Bound," 69, 71, except the quotation from another Mountaineer, *Times*, August 15, 1907. See also Banks, "Mountaineers," 79.
10. *Times*, August 3, 1907.
11. Ibid.; and Banks, "Mountaineers," 79.
12. [Curtis], "The Ascent of Mt. Olympus," *Mountaineer* 1 (March 1907): 20–24; and [Curtis], "Mountaineers' Annual Outing," 45–47.
13. Banks, "Mountaineers," 75.
14. Minutes, September 9, 1907, vol. 1, microfilm, ACP2, UW.
15. Curtis to E. B. Webster, December 28, 1915, box 1, file 28, ACP, UW.
16. Thomas R. Cox, *The Lumberman's Frontier: Three Centuries of Land Use, Society, and Change in America's Forests* (Corvallis: Oregon State University Press, 2010), 263–330.
17. Ibid., 331–75.
18. Curtis to Quilcene Improvement Club, January 31, 1912, box 1, file 18, ACP, UW.
19. Curtis to Webster, n. 15.
20. Rudo L. Fromme, "Fromme's Olympic (Forest) Memoirs, 1912 to 1916 [sic]." Rudo L. Fromme Papers, accession 1999-2 (RFP), UW, 62 (quotation), 63.
21. Olson, first in series of articles on the trip, n.d., *Aberdeen Daily World* (*World*), box 18, Scrapbook Relating to Aberdeen and the Olympic Mountains (AOMS), ACP, UW.
22. Unattributed clipping in AOMS, ACP, UW.
23. Olson, n. 21.
24. Olson, third article, n.d., *World*, AOMS, ACP, UW.
25. Olson, ninth article, December 6, 1924, *World*, AOMS, ACP, UW.
26. Olson, tenth article (attributed), December 6, 1924, *World*, AOMS, ACP, UW.
27. Fromme, "Fromme's Memoirs," RFP, UW, 63–64.
28. Ibid., 63.
29. Curtis article, number not indicated, November 12, 1924 (attributed), *World*, AOMS, ACP, UW.
30. Curtis draft article, box 1, file 4, Speeches and Writings, accession 4058-3, ACP, UW.
31. Curtis to Sawyer, September 30, 1929, box 6, file 24, ACP, UW.
32. Ibid.
33. Ibid.

Notes to Chapter Ten

1. Curtis to Ernest Walker Sawyer, November 15, 1932, box 10, file 1, ACP, UW.
2. T. H. Watkins, *Righteous Pilgrim: The Life and Times of Harold L. Ickes, 1874–1952* (New York: Henry Holt, 1990), 551–52.

3. Curtis to Marshall N. Dana, August 17, 1934, box 10, file 34, ACP, UW. Curtis thought of the wilderness enthusiasts of the 1930s as silly rather than elitist, so it would be invalid to connect his thinking directly to later criticisms of the more influential wilderness proponents. The atmosphere of the 1960s wilderness movement nonetheless influences thinking about Curtis, so some observations are in order. The charge of elitism in an age of superficial egalitarianism makes defenders of the 1960s wilderness proponents uneasy; see Turner, *Promise of Wilderness.* Turner denies the special interests of the cultural-intellectual elite who favored wilderness by stating that others presumably not of the elite, "hunters, fishermen, and other outdoor enthusiasts in the West," were supporters of wilderness, and that wilderness was accessible to all, p. 31. The accessibility of wilderness is demonstrated first, by assertion, and then by the statement that in 1965 visits to wilderness reached "3.5 million visitor days per year" (p. 31), but with no analysis of visitor location within any wilderness area, or length of stay, or of repeat visits, or of any comparison with visits to natural areas not designated as wilderness. Finally, he defends wilderness on the grounds that its advocates feel better because it exists, a psychic benefit at taxpayer expense unavailable to the unreflective masses. On pp. 67–68, Turner again denies the charge of elitism with the observation that no wilderness buffs were wealthy, a confusion of two types of elitism, his elite of culture and intellect with an elite of wealth.
4. Lien, *Olympic Battleground*, 119–21; and Irving Brant, *Adventures in Conservation with Franklin D. Roosevelt* (Flagstaff, AZ: Northland, 1988), 33–36, 55, 74, 116.
5. Dyana Z. Furmansky, *Rosalie Edge, Hawk of Mercy: The Activist Who Saved Nature from the Conservationists* (Athens: University of Georgia Press, 2010), 3, 88, 109.
6. Ibid., 1–6, 148–49, 207–08.
7. Ibid., 116–18; and Brant, *Adventures in Conservation*, 97, 116 (quotation).
8. R. L. Fromme, quoted in Twight, *Organizational Values*, 39; and Curtis to J. J. Underwood, December 8, 1924, box 2, file 28, ACP, UW.
9. David H. Madsen to Preston P. Macy, September 5, 1934, box 1, file 2, Preston P. Macy Papers, accession 3211-1 (PMP), UW. See also Elmo R. Richardson, "Olympic National Park: Twenty Years of Controversy," *Forest History* 12 (April 1968): 8; Curtis to F. W. Mathias, June 23, 1934, file 31; and Curtis to Arno B. Cammerer, August 16, 1934, file 34, both in box 10, ACP, UW.
10. Curtis to George Welch, December 30, 1935, box 11, file 17. ACP, UW; Lien, *Olympic Battleground,* 129–31; and Catton, *National Park, City Playground*, 83–105.
11. House Committee on Public Lands, *Mount Olympus National Park: Hearing[s] before the Committee on the Public Lands*, 77th Cong; 2d sess., 1936, pp. 167–69; and Curtis to Harold B. Say, April 20, 1935 (quotations), box 11, file 9, ACP, UW.
12. Curtis to Mon Wallgren, April 8, 1935, box 11, file 8, ACP, UW.
13. Ibid.
14. Curtis to Cammerer, April 27, 1935 (first quotation), file 9; and Curtis to Cammerer, September 12, 1935 (second quotation), file 14, both in box 11, ACP, UW.
15. Wallgren to Curtis, in *Mount Olympus National Park*, 169.
16. Curtis, 167.
17. Ibid., 169.
18. Yeon, "The Issue of the Olympics," *American Forests* 42 (June 1936): 255–57, 291–92. For quotations, see 255, 291. For Yeon, see Gideon Bosker and Lena Lencek, *Frozen Music: A History of Portland Architecture* (Portland, OR: Western Imprints, 1985), 89–90, 97, 112–16.
19. *Mount Olympus National Park*, 160–62, 163 (first quotation), 164–78, 179 (second quotation).
20. Ibid., 160–79.
21. Curtis to L. G. McClelland, May 16, 1936, box 11, file 22, ACP, UW.

22. *Mount Olympus National Park*, 163, 173, 179.
23. Ibid., 160 (first quotation), 169 (second quotation), 170 (third quotation).
24. Wallgren to Curtis, April 16, 1935, in ibid., 169.
25. Cammerer to Curtis, May 4, 1935, in ibid., 169–70 (quotation); and Cammerer to Curtis, November 20, 1935, in ibid. Cammerer to Curtis, October 23, 1935, box 11, file 15, ACP, UW.
26. Curtis to Martin, May 14, 1936, box 11, file 21, ACP, UW.
27. Curtis to Pearl Wanamaker, July 18, 1936, ibid., file 24.
28. *Mount Olympus National Park*, 163–69, 179 (quotation).
29. Richardson, "Olympic National Park," 9, argues for a prior decision.
30. Curtis to McClelland, n. 21.
31. Curtis to Cammerer, June 27, 1936 (first quotation), file 23; and Cammerer to Curtis, July 1, 1936 (second quotation), file 24, both in box 11, ACP, UW.
32. Richardson, "Olympic National Park," 8, 9 (quotation); Cammerer to Mrs. E. B. Ackerman, n.d., box 1, file 20, PMP, UW; Twight, *Organizational Values*, 87; and *New York Herald Tribune*, March 19, 1937.
33. Curtis to Cammerer, March 1, 1937, box 11, file 29, ACP, UW.
34. Curtis to Martin, n. 26.
35. Richardson, "Olympic National Park," 9.
36. Ibid.
37. Owen A. Tomlinson, "Confidential Memorandum for the Files: Regarding President Roosevelt's Visit to the Olympic Peninsula, September 30–October 1, 1937" (quotations); Preston P. Macy, memorandum, October 5, 1937, both in box 1, file 22, PMP, UW; see also n. 38.
38. Benj. N. Phillips, to Newton B. Drury, May 18, 1943, copy, box 1, file 14, PMP, UW rebuts the "children's crusade" claims in Furmansky, *Rosalie Edge*, 214; McNulty, *Olympic National Park;* 260; and Watkins, *Righteous Pilgrim*, 566–67 (quotation).
39. Roosevelt indirectly quoted in Tomlinson, "Confidential Memorandum," n. 37.
40. Ibid.; for Ford and farming-industry, see Reynold M. Wik, *Henry Ford and Grass-roots America* (Ann Arbor: University of Michigan Press, 1972), 105–25, 142–61, although Wik emphasizes the farming aspect. See also Gregg Grandin, *Fordlandia: The Rise and Fall of Henry Ford's Forgotten Jungle City* (New York: Picador, 2009), 58–65, 256–63. For Wright, see Robert C. Twombly, *Frank Lloyd Wright: An Interpretive Biography* (New York: Harper & Row, 1973), 176–85, 239. A wry look at a Roosevelt farming-industry attempt is C. J. Mahoney, *Back to the Land: Arthurdale, FDR's New Deal, and the Costs of Economic Planning* (Hoboken, NJ: Wiley, 2011).
41. Roosevelt directly quoted in Tomlinson, "Confidential Memorandum," n. 37.
42. Roosevelt indirectly quoted in ibid.
43. Ibid.
44. Ibid.
45. Watkins, *Righteous Pilgrim*, 567 (quotation); and Lien, *Olympic Battleground*, 178.
46. Twight, *Organizational Values*, 99; Watkins, *Righteous Pilgrim*, 567–68; Richardson, "Olympic National Park," 10–14; and Lien, *Olympic Battleground*, 184–207.
47. House Committee on Public Lands, *To Establish the Olympic National Park in the State of Washington: Hearings before the Committee on Public Lands*, 75th Cong, 3d sess., 1938, pp. 30–31.
48. Brant, *Adventures in Conservation*, 221–24; and Watkins, *Righteous Pilgrim*, 568. Curtis joined a short-lived group, the Washington State Resources Federation, opposed to expansion of the Olympic National Park, as one of six vice presidents. The federation could do little about the Olympic park because its expansion following the 1938 legisla-

tion occurred either by law or by federal land purchases. The federation's opposition to a national park in the North Cascade Mountains was successful over the short term. See Chapter Twelve and *Times,* November 30, 1939; December 1, 1939; and January 3, 1939.

49. Curtis to F. W. Mathias, July 13, 1938, box 11, file 35, ACP, UW.
50. Curtis to Raymond L. Hornbeck, January 19, 1940, box 12, file 6, ACP, UW.
51. Curtis to F. W. Mathias, January 9, 1940, box 12, file 6, ACP, UW.
52. Curtis to Will H. Taylor, November 22, 1924, box 2, file 27, ACP, UW.
53. Curtis to A. H. Vandenberg, February 26, 1940, box 12, file 7, ACP, UW (first quotation). On the issue of resident removal to develop national parks, see Paul S. Sutter, *Driven Wild: How the Fight against Automobiles Launched the Modern Wilderness Movement* (Seattle: University of Washington Press, 2002), 12, 266 n. 19. For Curtis on Ickes, see Curtis to Tim Healy, July 25, 1940; Albert C. Martin to Curtis, July 26, 1940, both in box 12, file 11, ACP, UW; and Curtis to Charles L. McNary, February 16, 1940, box 12, file 7, ACP, UW (second quotation).
54. Curtis to W. W. Washburn Jr., July 5, 1940, box 12, file 11, ACP, UW.
55. Richardson, "Olympic National Park," 14.
56. Curtis to Roosevelt, November 28, 1940, box 12, file 13, ACP, UW.
57. Curtis to Mathias, n. 51.
58. Curtis to Alex Polson, June 20, 1933, box 10, file 10, ACP, UW.
59. Curtis to H. H. Van Brocklin, July 28, 1933; and Van Brocklin to Curtis, July 14, 1933, with Curtis note on back of second page of Van Brocklin letter about revised route, both in box 10, file 11, ACP, UW. For quotation, see T. J. Remann to Curtis, July 10, 1933, box 10, file 11, ACP, UW.
60. Curtis to T. A. Stevenson, January 15, 1937, box 11, file 28, ACP, UW (quotation); and excerpt from hearings, box 1, file 41, PMP, UW.
61. Richardson, "Olympic National Park," 14.
62. Sutter, *Driven Wild,* 257. Turner, in *Promise of Wilderness,* is conscious of the ironies involved in lightweight equipment manufacturers, representatives of the corporate system so distrusted (to say the least) by the pro-wilderness left, making possible the easy penetration of wilderness areas theretofore the province of relatively few. He takes refuge in the belief that lightweight gear will enable the new masses of wilderness visitors to travel with little or no trace through terrain previously unspoiled, pp. 79–80, 91–94.
63. Shankland, *Steve Mather,* 42–43, 82, 172.
64. Lien, *Olympic Battleground,* 44.
65. Brant, *Adventures in Conservation,* 118.
66. For farming in the corridor, see Evans, *Historic Resource Study,* 94–95, 113; and Thomas T. Aldwell, *Conquering the Last Frontier* (Seattle: Artcraft, 1950), 45–46.

Notes to Chapter Eleven

1. Land and Water Contract, Grandview Orchard Tracts, January 8, 1907. The purchase price was $1,592.50, with one-third down and the balance, plus interest, in two annual installments. Records of the Sunnyside Valley Irrigation District, Sunnyside, WA (SVID). Almost three years later Curtis won five acres of land along the Sunnyside canal in a contest sponsored by the Washington Irrigation Company. Entrants photographed the conversion of raw desert land into a productive irrigated farm, which Curtis best accomplished in a series of ten "artistically colored" photos. The five-acre tract was not adjacent to the ranch or it would have been incorporated into that property or mentioned in connection with Curtis's original holding. It was not, nor is it mentioned in any surviving Curtis materials. The probability is that Curtis never took title to the property. How he disposed of his prize is unknown. See *Times,* November 28, 1909.

2. Curtis, "Address of Asahel Curtis at Grade and Pack Conference," Spokane, March 1927, box 1, file 5, ACP3, UW; Roscoe Sheller, *Courage and Water: A Story of Yakima Valley's Sunnyside* (Portland, OR: Binfords and Mort, 1952), 13, 57, 81, 118; unattributed obituary of Blaine, Biography file Bla–Blai, UW; and "Elbert F. Blaine," in Bagley, *History of Seattle*, vol. 2, pp. 796–98.
3. For water rights and general information on reclamation and irrigation early in Curtis's era, see Frederick Haynes Newell and Daniel William Murphy, *Principles of Irrigation Engineering* (New York: McGraw-Hill, 1913), 263–74. All information not specifically cited to other sources is from this book. See also Donald J. Pisani, *Water, Land, and Law in the West: The Limits of Public Policy, 1850–1920* (Lawrence: University Press of Kansas, 1996), 10.
4. Newell and Murphy, *Principles Irrigation Engineering*, 138, and Plate Y, fig. A. See also Frederick Haynes Newell, *Water Resources: Present and Future* (New Haven: Yale University Press, 1920), 237.
5. A similar episode is in C. Brewster Coulter, "The New Settlers on the Yakima Project, 1880–1914," *PNQ* 61 (January 1970): 14.
6. For 1901, see Calvin B. Coulter, "The Victory of National Irrigation in the Yakima Valley, 1902–1906," *PNQ* 42 (April 1951): 108; and for 1903–05, see Coulter, "New Settlers," 10.
7. Coulter, "New Settlers," 14.
8. *The Sunnyside Story* (Sunnyside, WA: Sunnyside Museum and Historical Association, 1982), 5 (quotation); Coulter, "New Settlers," 14; and Sheller, *Courage and Water*, 119–21, 123.
9. Coulter, "Victory," 106–08; Pisani, *Water, Land, and Law*, 229 n. 19; and Donald Worster, *Rivers of Empire: Water, Aridity, and the Growth of the American West* (New York: Oxford University Press, 1985).
10. Coulter, "Victory," 102.
11. Histories of reclamation not cited above include Donald J. Pisani, *To Reclaim a Divided West: Water, Law and Public Policy, 1848–1902* (Albuquerque: University of New Mexico Press, 1992); William D. Rowley, *Reclaiming the Arid West: The Career of Francis G. Newlands* (Bloomington: Indiana University Press, 1996); James R. Kluger, *Turning on Water With a Shovel: The Career of Elwood Mead* (Albuquerque: University of New Mexico Press, 1992); Kathryn L. Utter, "In the End the Land: Settlement of the Columbia Basin Project." Ph.D. diss., University of Washington, 2004, especially strong in the resettlement programs of the 1920s and 1930s, and settlement operations; and Lawrence J. MacDonnell, *From Reclamation to Sustainability: Water, Agriculture, and the Environment in the American West* (Niwot: University Press of Colorado, 1999).
12. Washington State Planning Council (WSPC), *Reclamation, A Sound National Policy: An Inquiry into the Effects of Irrigation Development on Local, State and National Economy As Demonstrated by The Yakima Valley And Other Irrigated Areas In Washington* ([Olympia]: WSPC, 1936), 11.
13. Bayard Still, *Urban America: A History with Documents* (Boston: Little, Brown, 1974), 79, table 2.2; and 364, table 3.3.
14. Blake McKelvey, *American Urbanization: A Comparative History* (Glenview, IL: Scott, Foresman, 1973), 73, table 6.
15. Ibid., 104, table 8.
16. Howard P. Chudacoff and Judith E. Smith, *The Evolution of American Urban Society*, 4th ed. (Englewood Cliffs, NJ: Prentice Hall, 1994), 112, table 4-1.
17. Ibid., 115–19, 144–45; and David R. Goldfield and Blaine A. Brownell, *Urban America: A History*, 2nd ed. (Boston: Houghton Mifflin, 1990), 250–59; and Pisani, *Water, Land, and Law*, 138.

18. Pisani, *Water, Land, and Law*, 138 (quotation). For the closing frontier and the "frontier thesis" as shaping American character, see Allen G. Bogue, *Frederick Jackson Turner: Strange Roads Going Down* (Norman: University of Oklahoma Press, 1998), 91–118.
19. Coulter, "Victory," 110–15; Sheller, *Courage and Water*, 118, 122–24; and Timothy A. Dick, *The Yakima Project* (Denver: Bureau of Reclamation History Program, 1993), 10.
20. Betty McCullough to Archie Satterfield, September 20, 1969, box 3, file Curtis Mountains, AS, APRCA, UAF; G. Thomas Edwards, "Irrigation in Eastern Washington, 1908–1911: The Promotional Photographs of Asahel Curtis," *PNQ* 72 (July 1981), 112–20; G. Thomas Edwards, "'The Early Morning of Yakima's Day of Greatness': The Yakima County Agricultural Boom of 1905–1911," *PNQ* 73 (April 1982), 78–89; and Mark Nielsen, "The Brown Farm on the Nisqually Delta, 1904–1919: A Photographic Essay," *PNQ* 71 (October 1980), 162–71. For Curtis's interest in land settlement and farm prosperity whether or not related to reclamation, see *Times*, January 19, 1928; February 5, 1928; and July 24, 1929.
21. Edwards, "Early Morning of Yakima's Day," 87.
22. Coulter, "New Settlers," 15.
23. Ibid., 15 ("clerks" quotation), 17 (second quotation), 18–21.
24. Photographs no. 16457, year 1906, pp. 1, 191; and no. 20741, April 1910, p. 23, album 137, ACC, WSHS.
25. Betty McCullough to Archie Satterfield, April 11 [1969], box 3, file Curtis Mountains, AS, APRCA, UAF.
26. Photograph no. 42398, no album, ACC, WSHS.
27. McCullough to Satterfield, n. 25.
28. Ibid.
29. Ibid.
30. For the size of farms in 1935, see WSPC, *Reclamation*, 42. Payroll and crop photographs are in album 137 and elsewhere in ACC, WSHS. See also McCullough to Satterfield, n. 25.
31. For apples, see www.orangepippin.com/apples/winesap and www.orangepippin.com/apples/newtown-pippin; and McCullough to Satterfield, n. 25.
32. McCullough to Satterfield, n. 25.
33. Photograph no. 42400, October 18, 1921, album 137; and photograph no. 42437, no album, ACC, WSHS. For Ryther, see Mildred Andrews, "Ryther, Mother Olive (1849–1934), "Essay 546, History Link, www.historylink.org/index.cfm?DisplayPage=output.cfm&file_id=546.
34. Photographs no. 35474 and no. 33478 (Betty), both in album 137, ACC, WSHS.
35. McCullough to Satterfield, n. 25.
36. Curtis to Frank H. Lamb, May 19, 1928, box 5, file 14, ACP, UW.
37. Consecutive tax records for Curtis's property are in Real Property Assessment, Lands, Yakima, Washington Book P. A. 2; Real Property, Assessment, Lands, Yakima, Washington; and Real Property Assessment, Lands, Yakima, Washington Book PA-1, Central Region Branch, Washington State Archives, Ellensburg. Records of the foreclosure, transfer of sale document, and assumption of title are in C. D. Stephens, Treasurer, Yakima County, to Perham Fruit Company, fee no. 712722, courtesy of the Fidelity Title Company, Yakima, WA. The mortgage is Mortgage, Asahel Curtis *et. Ux.* to Grandview Cold Storage, March 23, 1935, fee no. 685190, Office of the Assessor, Yakima County, WA. For the relationship between the Grandview Cold Storage Company and the Perham Fruit Company, see John Baule, "Last Bite [Ben Perham]," *Good Fruit Grower*, January 1, 2010, online archive, www.goodfruit.com/last-bite-13. The steps in the arrangement outlined in the government documents could not have occurred without advance agreement

among the principals, although the entire plan probably unfolded piecemeal. The cold storage company's paying the taxes on Curtis's property could not have happened without agreement. The public sale realized $12.84, not a large sum even in 1934. The sale was to Donald L. Allen, probably an employee of Perham Fruit, who the following month assigned the certificate of sale to the Perham firm. Benjamin A. Perham, the company's founder and president, was an important figure in eastern Washington. Perhaps he had made arrangements to bid on the property through an intermediary (Donald L. Allen), which would explain the low price and apparent lack of competing bids. Others might have desired the Curtis land and buildings, but also might have wished to continue pleasant relationships with Grandview Cold Storage and Perham Fruit. Perham was president of both companies and of others besides. The assertion of Asahel and Florence Curtis in their mortgage document that they had "a valid and unincumbered title in fee simple to the said premises" would have been false to anyone who knew of the conditional sale of the property, which would include anyone at Grandview Cold Storage authorized to accept the mortgage. The mortgage amount was $1,846.15, a precise figure indicating that most of it would not go to Curtis, but to repay taxes and other costs of carrying the Curtis property from 1931. Betty's memory of "a couple of hundred dollars" paid to her father could have been a payment arranged in advance. The final dollar figure might not have been known when the mortgage was prepared, but the residue of the mortgage amount could have been sent to Curtis as an informal acknowledgement of his booster activities and other work for irrigation and reclamation in the Yakima Valley. When no one redeemed Curtis's ranch and Perham Fruit assumed title, it was the apparent end of the matter. Curtis's dismay on receiving the check could have been partly because he expected a larger amount than the one he obtained.

38. McCullough to Satterfield, n. 25.
39. Robert Wayne Wilson, "A History of the Washington State Apple Advertising Commission" (master's thesis, University of Washington, 1966), 60–120. For expositions, see "Pacific Northwest Fruit Exposition: an important Event in the Fruit History of the Northwest," *Western Fruit Jobber* 8 (December 1921): 44–47; *Times,* November 26, 1921; and *Times,* October 29, 1922. For packing and grading, see Wilson, above, 30–39; and www.bestapples.com/facts/facts_grades.aspx. For apple shows, especially the Spokane International Apple Show, see Wilson, above, 39–45. For an exhibition car, see Wilson, above, 45–49.
40. Wilson, "History Washington Apple Commission," 60–61; *Times,* March 26, 1926, April 25, 1926, and March 18, 1927; Curtis to Fred C. Neely, October 24, 1927, box 4, file 29, ACP, UW; and Botsford-Constantine Company, Advertising, *Pacific Northwest Boxed Apples* (Seattle, 1926), 10, 20 (quotation).
41. Curtis to A. S. F. Steele, April 20, 1927, box 4, file 14; Curtis to Frank H. Lamb, April 21, 1927, box 4, file 14 (quotations); and Curtis to Neely, n. 40, all in ACP, UW.
42. For the founding, see *Proceedings, Second Annual Meeting of the Washington Irrigation Institute, Yakima, Washington December 16, 17, 18, 1914* (North Yakima, WA: Washington Irrigation Institute [WII], 1915), 7–9, UW. For Curtis's participation, see *Proceedings, Eighth Annual Meeting, Ellensburg, December 16–17, 1920* (Yakima, 1921?), 58 (quotation); and for membership, 106. For Coulter quotation, see Coulter, "New Settlers," 17.
43. *Times*, November 16, 1928.
44. *Yakima Daily Republic*, November 7, 1929. See also Minutes of the Seventeenth Annual Meeting, Yakima, November 7 & 8, 1929, box 1, file 2, Records of the WII, 1927–1932, accession 4952-1 (WII), UW.
45. *Yakima Daily Republic*, November 8, 1929.
46. Transcript, box 1, file 2, WII, UW.

47. *Proceedings, Twentieth Annual Meeting, Sunnyside, November 19, 1932*, and *Twenty-first Annual Meeting, Yakima, November 24–25, 1933.*
48. Bureau of Reclamation, Elwood Mead to Curtis, November 1, 1929, box 6, file 27, ACP, UW. Boxed Apple Bureau, photograph no. 58915 and others, album 134; Kittitas, no. 62137, album 124; and Roza, no. 60273, album 104, all in ACC, WSHS.
49. Robert M. Carriker, *Urban Farming in the West: A New Deal Experiment in Subsistence Homesteads* (Tucson: University of Arizona Press, 2010), 10. Problems and opportunities associated with reclamation are in WSPC, *Reclamation*, 9–10; See also A. S. Goss to Curtis, November 1, 1932, box 9, file 37, ACP, UW; and J. J. Underwood to Curtis, November 14, 1932, box 10, file 2, ACP, UW.
50. WSPC, *Reclamation*, 4 (all quotations except Curtis's), 5. For the Curtis-Mead cordiality, see Mead to Curtis, n. 48. Mead's problems are recounted in MacDonnell, *Reclamation*, 3–113; and Kluger, *Turning on Water*, 51–73, 78–107, 155. The Curtis quotation "his... Kittitas" is in Curtis to A. L. B. Davis, December 11, 1929, box 6, file 23, ACP, UW. See also *Times*, July 8, 1928; and December 11, 1928.
51. WSPC, *Reclamation*, 47, tables I and II; and 48, tables III and IV.
52. Ibid., 101, table "Summary of Apple Exports from Pacific Coast;" and conclusions on 52–53.
53. On special financing, see Curtis to Fred W. Graham, April 5, 1930, box 7, file 19; and Curtis to members of the reclamation committee of the Washington State Chamber of Commerce, July 26, 1932, box 9, file 33, both in ACP, UW. See also the exchange of letters between Curtis and James O'Sullivan, for example, Curtis to O'Sullivan, August 17, 1934, and O'Sullivan to Curtis, August 29, 1934, both in box 10, file 34, ACP, UW.
54. Paul Pitzer, *Grand Coulee: Harnessing a Dream* (Pullman: Washington State University Press, 1994), 1–6, 9–80. For Curtis's standing in the Columbia Basin Irrigation League, see Roy R. Gill to Curtis, September 15, 1927, letterhead, box 4, file 26, ACP, UW. For Some of Curtis's activities in the league see *Times*, May 15, 1927; October 4, 1927; and October 22, 1929.
55. Curtis to Roy R. Gill, June 25, 1928 (first quotation), box 5, file 16; and Curtis to Christy Thomas, May 18, 1927, box 4, file 17, both in ACP, UW. *Spokane Spokesman-Review*, August 20, 1926; and *Times*, August 27, 1926. For the photograph see Curtis to Roy R. Gill, September 27, 1926, box 3, file 34, ACP, UW.
56. Curtis to Ralph D. Nichols, August 18, 1933 (quotation), box 10, file 4, ACP, UW. *Times*, April 7, 1935. For the commission, see Curtis to F. O. Hagie, January 11, 1937, box 11, file 28, ACP, UW; Kathleen Waugh, comp., *Roll on Columbia: Guide to the Records of the Columbia Basin Survey Commission and the Columbia Basin Commission, 1919–1964* (WSA, 1987), 7–27, WSA; and Utter, "In the End the Land," 103–238.
57. Curtis to James C. Bettis, August 29, 1938, box 89, file 3 (first quotation); minutes of the Columbia Gorge Committee, October 29, 1938, box 89, file 90/2-20-1 (second quotation); Asahel Curtis, Pierre J. Landry, and Donald T. Stewart, to the House and Senate rules committees, February 18, 1939, box 89, file 3; P. Hetherton to Curtis, October 21, 1940, box 89, file 4; First Report, The Honorable Earl Snell to the Legislatures of Oregon and Washington, December 1942, box 90; and other items in box 89, all in State Planning Council–Research and Development–Recreation–Columbia Gorge, 90/2-20-1, Records of the WSPC, Record Group 90 (RG 90), WSA.
58. B. H. Kizer to P. Hetherton, May 1, 1940, box 89, file 4, RG 90, WSA.
59. Minutes of the Public Works Committee of the WSPC, November 15, 1935, box 83, file 20/2-19, minutes no. 1, RG 90, WSA.
60. WSPC, *The Proposed Mount Olympus National Park: A Land Use Study of Public Lands On the Olympic Peninsula* (Olympia, WA, December 1936), box 11, RG 90, WSA; and Curtis

to Clarence D. Martin, March 7, 1938, box 74, file 90/2-13, Olympic Peninsula Correspondence 1938 no. 1, RG 90, WSA.

Notes to Chapter Twelve

1. Betty McCullough to Archie Satterfield, October 25 [1969], box 3, file Curtis Mountains, AS, APRCA, UAF.
2. Curtis to Gellatly, November 10, 1932; and to J. C. Hubbell, November 10, 1932, both in box 9, file 37, ACP, UW.
3. For the world's fairs and local exhibits see McCullough to Satterfield, September 20, 1969, box 3, file Curtis Mountains, AS, APRCA, UAF; *Times,* May 6, 1933; and *Times,* October 23, 1938. For the king and queen of Siam, see *Times,* September 8, 1931. For a photo contest, see *Times,* August 28, 1938. For the Washington guide, see *Times,* October 21, 1941.
4. Leo A. Borah, "Washington, the Evergreen State: The Amazing Commonwealth of the Pacific Northwest Has Emerged from the Wilderness in a Span of Fifty Years," *National Geographic Magazine* 58 (February 1933), 131–46.
5. Franklin Fisher to Frank W. Guilbert, January 31, 1933, copy, box 10, file 5, ACP, UW.
6. Key letters are Franklin Fisher to Curtis, November 14, 1931, box 9, file 7; Fisher to Curtis, July 15, 1932 (quotation), box 9, file 32; Gerard F. Hubbard to Curtis, September 14, 1932, box 9, file 37; and Fisher to Curtis, October 27, 1932, box 9, file 37, all in ACP, UW.
7. See, for example, E. C. Leedy to Curtis, October 26, 1928, box 5, file 28, ACP, UW.
8. Newhall, *History of Photography*, 272, 276.
9. Curtis to George Lawler, May 11, 1933, box 10, file 8, ACP, UW.
10. McCullough to Satterfield, n. 1. For Strong, see box T-151, untitled folder, CBR, WSHS.
11. De Steiguier to Curtis, May 4, 1938, box T-151, file Curtis, Asahel, correspondence regarding buying and selling photographs, SC, WSHS. The volumes were integrated into ACC. See, for example, "Alaska and Yukon Territory," vol. 1, ACC, WSHS.
12. Harriet Holmberg Williams to Jeanne [Engerman] and Richard [Frederick], October 8, 1982, box 314.2-28, no file, SC, WSHS. Lindsley, *Diary*, December 16, 29, 1930; January 3, February 26–27; and August 1, 1931, LLP, UW.
13. See Murrow's reply to Curtis, January 11, 1934, box 10, file 22, ACP, UW.
14. David W. Jones, *Mass Motorization + Mass Transit* (Bloomington: Indiana University Press, 2008), 70–93, summarizes the New Deal involvement.
15. L. I. Hewes, "A Quarter Century of Western Road Building," WSGRA, *Proceedings* Seattle, September 29–October 1, 1938.
16. One older but still informative place to begin reading about the progressive movement is Michael McGerr, *A Fierce Discontent: The Rise and Fall of the Progressive Movement in America, 1870–1920* (New York: Free Press, 2003). For Curtis's allegiance to Theodore Roosevelt and Herbert Hoover, see McCullough to Satterfield, September 20, 1969; and McCullough to Satterfield, October 25, [1969], box 3, file Curtis Mountains, AS, APRCA, UAF.
17. For PWA work, see Linda Flint McClelland, *Building the National Parks: Historic Landscape Design and Construction* (Baltimore: Johns Hopkins University Press, 1998), 332. For CCC work, see McClelland, *Building the National Parks*, 336–50; and Catton, *National Park, City Playground*, 99–103. For Tomlinson's directive, see McClelland, 340.
18. For Tomlinson, see RG 79, box 7, file H18, Tomlinson, Owen A.; I–K, file 36; U–Z, file 60; and box 42, file K3415, Artificial History Collection, PARS, NARA. See also Lauren Danner, "Ice Peaks National Park: The Spectacular Failure of a New Deal Idea," *Columbia* 23 (Fall 2009), 29–35.
19. Curtis to W. G. Oves, November 13, 1939, box 12, file 3; and to Miller Freeman, November 28, 1940, box 12, file 5, ACP, UW.

20. McCullough to Satterfield, n. 3.
21. McCullough to Satterfield, n. 1; For Walter's residence, *Polk Directory*, 1940, p. 398.
22. For the children's activities, various clippings, AS, APRCA, UAF; for Polly, Polly Kella obituary, community.seattletimes.nwsource.com/archive. For quotations, see box 314.2-28, file Curtis–Interview Notes, CBR, WSHS.
23. McCullough to Satterfield, n. 1.
24. Undated clipping, box 3, file Curtis Mountains, AS, APRCA, UAF.
25. *P-I* and *Times*, March 8; and *P-I*, March 10.
26. *Seattle Business* 24 (March 13, 1941): 2; *Washington Motorist* 22 (April 1941): 10; and WSGRA, *Proceedings*, Wenatchee, September 18-20, 1941.
27. For the safety council see *Times*, September 1, 1936, and April 3, 1938. For the streetcar accident, see unattributed clipping, "Photographer, Hit By Automobile, Hurt," box 3, file Curtis Correspondence, AS, APRCA, UAF. For other accidents, see *Times*, December 26, 1904. For the bridge groundbreaking, see *Times*, December 27, 1938. For the bridge dedication, see *Times*, June 5, 1940. For the Elks, see *Times*, July 31, 1938.
28. C. E. Johns, memorandum to Mrs. Lewis, October 23, 1941, box 24-1-18, folder Progress Commission, 1941, Subject Files, Arthur B. Langley Papers, WSA; and *Times*, March 31, 1974, for Florence's death notice.
29. Johns memorandum to Lewis, n. 28; and box Asahel Curtis–Articles of Incorporation, Account Books–Account books & Ledgers, Loose Inventory Sheets, especially Ledger 150, Ledger 300 18-R, and check stubs, CBR, WSHS.
30. Unattributed clipping, "C. B. Reed Is Charged with Theft," box 3, file Curtis Correspondence, AS, APRCA, UAF; *Times*, September 2, 1943 and November 4, 1943; and copy of minutes, August 22, 1943, box 314.2-28, CBR, WSHS. For one estimate of the number of Curtis photos in the WSHS, see *Times*, January 5, 1947.
31. McCullough to Satterfield, n. 1; and the precise dates in Satterfield, *Seattle*, 128, which appear to be accurate. For information about the Curtis memorial area, see a separate folder on the 1964 nature trail dedication, SC, WSHS; and *Times*, June 10, 1942.
32. "In Way of Explanation," and "On the Dedication of a nature trail in the Asahel Curtis Memorial Grove," separate folder on the 1964 nature trail dedication, SC, WSHS.
33. Duncan, "Curtis, Asahel"; Macdonald, "Image Conscious," 16; and Satterfield, *Seattle*, 128. The author attended a tour of the memorial grove during which the lightning strike story was again repeated.
34. J. Herbert Stone to Catherine May, June 25, 1963; "Snoqualmie National Forest, Asahel Curtis Recreation Area"; and "Suggested Recreation Trips and Facilities, North Bend Ranger District"; and other material in separate folder on the nature trail dedication, n. 31.
35. Mrs. Roland Brewer, "Presidential Address"; "Program, Dedication, Asahel Curtis Nature Trail"; photographs; and other material in separate folder on the nature trail dedication, n. 31.
36. The papers were acquired "from the Asahel Curtis Studio in 1942," "Guide to the Asahel Curtis Papers: 1898, 1908-1941," UW.
37. Popular illustrated works citing or referencing the Curtis papers are Frederick and Engerman, *Asahel Curtis*; Satterfield, *Klondike Park*; Satterfield, *Seattle*; and Sucher, ed., *Asahel Curtis Sampler*. Scholarly studies include Catton, *National Park, City Playground*; Cox, *Park Builders*; Lien, *Olympic Battleground*; and Louter, *Windshield Wilderness*.
38. Clipping, *Tacoma News Tribune*, April 15, 1979, box T-151, file Curtis, Asahel, Jr. Correspondence, SC, WSHS.
39. See Sources for a full citation.

Sources

Manuscript Collections and Other Primary Sources

Alaska and Polar Regions Collections and Archives, Elmer E. Rasmuson Library, University of Alaska Fairbanks
    Archie Satterfield Papers
Center for Pacific Northwest Studies, Bellingham, Washington
    Minutes of the North End Improvement Council (Kitsap County, Washington)
    Records of the Washington State Good Roads Association
Fidelity Title Company, Yakima, Washington
    Records concerning Asahel and Florence Curtis's farm property
National Archives and Records Administration, Pacific Alaska Region, Seattle
    Twelfth Census of the United States, 1900, manuscript
    Thirteenth Census of the United States, 1910, manuscript
    Fourteenth Census of the United States, 1920, manuscript
    Fifteenth Census of the United States, 1930, manuscript
    Record Group 79, Records of Mount Rainier National Park
Office of the Assessor, Yakima County, Washington
    Asahel and Florence Curtis Mortgage, 1935
Research Center, Washington State Historical Society, Tacoma
    Asahel Curtis Business Records
    Asahel Curtis Photograph Collection
    Asahel Curtis material, Special Collections
Special Collections, University of Washington Libraries, Seattle
    Asahel Curtis Papers
    Biography File
    Rudo L. Fromme Papers
    Lawrence Denny Lindsley Papers
    Preston P. Macy Papers
    Edmond S. Meany Papers
    Eula Lee Merrill Papers
    The Mountaineers Club of Seattle Records
    Reginald H. Thomson Papers
    Washington Irrigation Institute Records
State Historical Society of Wisconsin, online
    Records of the Wisconsin Office of the Adjutant General
Sunnyside Valley Irrigation District
    Sunnyside Valley Irrigation District Records
Washington State Archives, Central Regional Branch, Ellensburg
    Property Assessment Records, Yakima County
Washington State Archives, Olympia
    Governor Hartley Subject Files

Records of the Department of Highways, 1909-1922
Department of Highways Subject File
Arthur B. Langlie Papers
Ernest Lister Papers
Records of the Washington State Planning Council
Washington State Archives, Puget Sound Region Branch, Bellevue
Property Records, Kitsap County, WA

Books

Aldwell, Thomas T. *Conquering the Last Frontier*. Seattle: Artcraft, 1950.

Andrews, Ralph W. *Curtis' Western Indians*. Seattle: Superior, 1962.

_______. *This Was Logging! Selected Photos of Darius Kinsey*. Seattle: Superior, 1954.

Armbruster, Kurt E. *Orphan Road: The Railroad Comes to Seattle, 1853-1911*. Pullman: Washington State University Press, 1999.

Bagley, Clarence B. *History of Seattle: From the Earliest Settlement to the Present Time*. 3 vols. Chicago: S. J. Clarke, 1916.

Bankston, Russell A. *The Klondike Nugget*. Caldwell, ID: Caxton, 1935.

Barcott, Bruce. *The Measure of a Mountain: Beauty and Terror on Mount Rainier*. Seattle: Sasquatch, 1997.

Belasco, Warren James. *Americans on the Road: From Autocamp to Motel, 1910-1945*. Baltimore: Johns Hopkins University Press, 1997.

Berger, Michael J. *The Devil Wagon in God's Country: The Automobile and Social Change in Rural America, 1893-1929*. Hamden, CT: Archon Books, 1979.

Berton, Pierre. *The Klondike Fever: The Life and Times of the Last Great Gold Rush*. New York: Knopf, 1958. Rev. ed. *Klondike: The Last Great Gold Rush. 1896-1899*. Toronto: McClelland and Stewart, 1972.

Bogue, Allen G. *Frederick Jackson Turner: Strange Roads Going Down*. Norman: University of Oklahoma Press, 1998.

Bohn, Dave, and Rodolfo Petscheck. *Kinsey, Photographer: A Half Century of Darius and Tabitha May Kinsey*. 3 vols. San Francisco: Prism Editions, 1978-84.

Bosker, Gideon, and Lena Leneck. *Frozen Music: A History of Portland Architecture*. Portland, OR: Western Imprints, 1985.

Botsford-Constantine Company, Advertising. *Pacific Brand Apples*. Seattle: The Company, 1926.

Bowers, William L. *The Country Life Movement in America, 1900-1920*. Port Washington, NY: National University Publications, Kennikat Press, 1974.

Brant, Irving. *Adventures in Conservation with Franklin D. Roosevelt*. Flagstaff, AZ: Northland, 1988.

Bromberg, Nicolette. *Picturing the Alaska-Yukon-Pacific Exposition: The Photography of Frank H. Nowell*. Seattle: University of Washington Press, 2009.

Bryce, Robert M. *Cook & Peary: The Polar Controversy Resolved*. Mechanicsburg, PA: Stackpole, 1997.

Bureau of the Census. *Historical Statistics of the United States: Colonial Times to 1970*. 2 vols. Washington, DC: 1975.

Carriker, Robert M. *Urban Farming in the West: A New Deal Experiment in Subsistence Homesteads*. Tucson: University of Arizona Press, 2010.

Catton, Theodore. *National Park, City Playground: Mount Rainier in the Twentieth Century*. Seattle: University of Washington Press, 2006.

Chudacoff, Howard P., and Judith E. Smith. *The Evolution of American Urban Society*. 4th ed. Englewood Cliffs, NJ: Prentice Hall, 1994.

Clark, Norman H. *Mill Town: A Social History of Everett, Washington*. Seattle: University of Washington Press, 1970.

Coates, K. S., and W. R. Morrison. *The Alaska Highway in World War II: The U. S. Army of Occupation in Canada's Northwest*. Norman: University of Oklahoma Press, 1992.

Cox, Thomas R. *The Lumberman's Frontier: Three Centuries of Land Use, Society, and Change in America's Forests*. Corvallis: Oregon State University Press, 2010.

_______. *The Park Builders: A History of State Parks in the Pacific Northwest*. Seattle: University of Washington Press, 1988.

Curtis, Asahel. *Rainier National Park*. [Seattle: 1921].

Dalby, Milton A. *The Sea Saga of Dynamite Johnny O'Brien*. Seattle: Lowman & Hanford, 1933.

Davis, Barbara A. *Edward S. Curtis: The Life and Times of a Shadow Catcher*. San Francisco: Chronicle Books, 1985.

Dick, Timothy A. *The Yakima Project*. Denver: Bureau of Reclamation History Program, 1993.

Egan, Timothy. *Short Nights of the Shadow Catcher: The Epic Life and Immortal Photographs of Edward Curtis*. Boston: Houghton Mifflin Harcourt, 2012.

Evans, Gail E. H. *Historic Resource Study, Olympic National Park, Washington*. Seattle: National Park Service, 1983.

Ficken, Robert E. *Lumber and Politics: The Career of Mark E. Reed*. Santa Cruz, CA: Forest History Society Inc., and Seattle: University of Washington Press, 1979.

Filley, Bette. *The Big Fact Book about Mount Rainier*. Issaquah, WA: Dunamis House, 1996.

Frederick, Richard, and Jeanne Engerman. *Asahel Curtis: Photographs of the Great Northwest*. Tacoma: Washington State Historical Society, 1983. Reprinted 1985, 1986.

Frykman, George A. *Seattle's Historian and Promoter: The Life of Edmond Stephan Meany*. Pullman: Washington State University Press, 1998.

Furmansky, Dyana Z. *Rosalie Edge, Hawk of Beauty: The Activist Who Saved Nature from the Conservationists*. Athens: University of Georgia Press, 2010.

Gidley, Mick. *Edward S. Curtis and the North American Indian, Incorporated*. Cambridge, England: Cambridge University Press, 1988. Paperbound ed., 2000.

Goddard, Stephen B., *Getting There: The Epic Struggle between Road and Rail in the American Century*. New York: Basic, 1994.

Goetzmann, William H., and Kay Sloan. *Looking Far North: The Harriman Expedition to Alaska, 1899*. Princeton: Princeton University Press, 1982. Paperbound ed., 1983.

Goldfield, David R., and Blaine A. Brownell. *Urban America: A History*. 2nd ed. Boston: Houghton Mifflin, 1990.

Goodwin, Frederick K., and Kay Redfield Jamison. *Manic Depressive Illness: Bipolar Disorder and Recurrent Depression*. 2nd ed. New York: Oxford University Press, 2007.

Gordon, Sarah H. *Passage to Union: How the Railroads Transformed American Life, 1829-1929*. Chicago: Ivan R. Dee, 1996.

Grandin, Greg, *Fordlandia: The Rise and Fall of Henry Ford's Forgotten Jungle City*. New York: Picador, 2009.

Green, Lewis. *The Gold Hustlers*. Anchorage, AK: Alaska Northwest, 1977.

Gutfreund, Owen D. *Twentieth-Century Sprawl: Highways and the Reshaping of the American Landscape*. New York: Oxford University Press, 2004.

Haines, Aubrey L. *Mountain Fever: Historic Conquests of Rainier*. Portland: Oregon Historical Society, 1962. Reprinted, Seattle: University of Washington Press, 1999. Citations are from the reprint edition.

Hartman, John P. *A Brief History of the Washington State Good Roads Association*. Walla Walla, WA: The Association, 1939.

Henderson, Bruce. *True North: Peary, Cook, and the Race to the Pole*. New York: Norton, 2008.

Hokanson, Drake. *The Lincoln Highway: Main Street across America*. 10th anniversary ed. Iowa City: University of Iowa Press, 1999.

Hunt, William R. *North of 53°: The Wild Days of the Alaska-Yukon Mining Frontier, 1870–1914*. New York: Macmillan, 1974.

_______. *To Stand at the Pole: The Dr. Cook–Admiral Peary North Pole Controversy*. New York: Stein and Day, 1981.

Irish, Kerry E. *Clarence C. Dill: The Life of a Western Politician*. Pullman: Washington State University Press, 2000.

Jones, David W. *Mass Motorization + Mass Transit*. Bloomington: Indiana University Press, 2008.

Kirk, Ruth. *Sunrise to Paradise: The Story of Mount Rainier*. Seattle: University of Washington Press, 1999.

Kjeldsen, Jim. *The Mountaineers: A History*. Seattle: Mountaineers, 1988.

Kluger, James R. *Turning on Water with a Shovel: The Career of Elwood Mead*. Albuquerque: University of New Mexico Press, 1992.

Lawlor, Laurie. *Shadow Catcher: The Life and Work of Edward S. Curtis*. New York: Walker, 1994. Reprint, Lincoln: University of Nebraska Press, 2005.

Lewis, Tom. *Divided Highways: Building the Interstate Highways, Transforming American Life*. New York: Viking Penguin, 1992. Rev. ed. Ithaca, NY: Cornell University Press, 2013.

Lien, Carston. *Olympic Battleground: The Power Politics of Timber Preservation*. 2nd ed. Seattle: Mountaineers, 2000.

Louter, David. *Windshield Wilderness: Cars, Roads, and Nature in Washington's National Parks*. Seattle: University of Washington Press, 2006.

MacDonnell, Lawrence J. *From Reclamation to Sustainability: Water, Agriculture, and the Environment in the American West*. Niwot: University Press of Colorado, 1999.

Mahoney, C. J. *Back to the Land: Arthurdale, FDR's New Deal, and the Costs of Economic Planning*. Hoboken, NJ: Wiley, 2011.

Makepeace, Ann. *Edward S. Curtis: Coming to Light*. Washington, DC: National Geographic, 2001.

McClelland, Linda Flint. *Building the National Parks: Historic Landscape Design and Construction*. Baltimore: Johns Hopkins University Press, 1998.

McGerr, Michael. *A Fierce Discontent: The Rise and Fall of the Progressive Movement in America, 1870–1920*. New York: Free Press, 2003.

McKelvey, Blake. *American Urbanization: A Comparative History.* Glenview, IL: Scott, Foresman, 1973.

McNulty, Tim. *Olympic National Park: A Natural History.* Seattle: University of Washington Press, 2003.

McShane, Clay. *Down the Asphalt Path: The Automobile and the American City.* New York: Columbia University Press, 1994.

Miller, Orlando. *The Frontier in Alaska and the Matanuska Colony.* New Haven: Yale University Press, 1975.

Moley, Raymond, Jr. *The American Legion Story.* New York: Duell, Sloan and Pearce, 1966.

Morgan, Murray. *One Man's Gold Rush: A Klondike Album.* Seattle: University of Washington Press, 1967.

Nash, Roderick Frazier. *Wilderness and the American Mind.* 4th ed. New Haven: Yale University Press, 2001.

Naske, Claus-M. *Paving Alaska's Trails: The Work of the Alaska Road Commission.* Lanham, MD: University Press of America, 1986.

Newell, Frederick Haynes. *Water Resources: Present and Future.* New Haven: Yale University Press, 1920.

Newell, Frederick Haynes, and Daniel William Murphy. *Principles of Irrigation Engineering.* New York: McGraw-Hill, 1913.

Newhall, Beaumont. *The History of Photography.* New York: Museum of Modern Art, 1982.

Patton, Phil. *Open Road: A Celebration of the American Highway.* New York: Simon & Schuster, 1986.

Pisani, Donald J. *To Reclaim a Divided West: Water, Land and Public Policy, 1848-1902.* Albuquerque: University of New Mexico Press, 1992.

_______. *Water, Land, and Law in the West: The Limits of Public Policy, 1850-1920.* Lawrence: University Press of Kansas, 1996.

Pitzer, Paul C. *Grand Coulee: Harnessing a Dream.* Pullman: Washington State University Press, 1994.

Raibmon, Paige. *Authentic Indians: Episodes of Encounter from the Late-Nineteenth-Century Northwest Coast.* Durham: Duke University Press.

Ridge, Alice A., and John Wm. Ridge. *Introducing the Yellowstone Trail: A Good Road from Plymouth Rock to Puget Sound.* Altoona, WI: Yellowstone Trail Publishers, 2000.

Rose, Mark H, Bruce E. Seely, and Paul F. Barrett. *The Best Transportation System in the World: Railroads, Trucks, Airlines, and American Public Policy in the Twentieth Century.* Philadelphia: University of Pennsylvania Press, 2010. First published in 2006 by Ohio State University Press.

Rowley, William D. *Reclaiming the Arid West: The Career of Francis G. Newlands.* Bloomington: Indiana University Press, 1996.

Runte, Alfred. *Trains of Discovery: Railroads and the Legacy of Our National Parks.* 5th ed. Lanham, MD: Roberts Rinehart, 2011.

Sale, Roger. *Seattle: Past to Present.* Seattle: University of Washington Press, 1976.

Satterfield, Archie. *Klondike Park: From Seattle to Dawson City.* Golden, CO: Fulcrum, 1993.

_______. *Seattle: An Asahel Curtis Portfolio.* San Francisco: Chronicle Books, 1985.

Scherer, Joanna Cohan. *Edward Sheriff Curtis.* London: Phaidon, 2008.
Schwantes, Carlos Arnaldo. *The Pacific Northwest: An Interpretive History.* rev. and enl. ed. Lincoln: University of Nebraska Press, 1996.
Seeley, Bruce E. *Building the American Highway System: Engineers as Policy Makers.* Philadelphia: Temple University Press, 1987.
Sellars, Richard West. *Preserving Nature in the National Parks: A History.* New Haven: Yale University Press, 1997.
Shankland, Robert. *Steve Mather of the National Parks.* 3d ed., rev. and enl. New York: Knopf, 1970.
Sheller, Roscoe. *Courage and Water: A Story of Yakima Valley's Sunnyside.* Portland, OR: Binfords and Mort, 1952.
Still, Bayard. *Urban America: A History with Documents.* Boston: Little, Brown, 1974.
Sucher, David, ed. *The Asahel Curtis Sampler: Photographs of Puget Sound Past.* Seattle: Puget Sound Access, 1973.
Sunnyside Museum and Historical Association. *The Sunnyside Story.* Sunnyside, WA: The Association, 1982.
Sutter, Paul S. *Driven Wild: How the Fight Against Automobiles Launched the Modern Wilderness Movement.* Seattle: University of Washington Press, 2002.
Swain, Donald C. *Wilderness Defender: Horace M. Albright and Conservation.* Chicago: University of Chicago Press, 1970.
Swift, Earl. *The Big Roads: The Untold Story of the Engineers, Visionaries, and Trailblazers Who Created the American Superhighways.* Boston: Houghton Mifflin Harcourt, 2011.
Tuhy, John E. *Sam Hill: The Prince of Castle Nowhere.* Beaverton, OR: Timber Press, 1983. Reprint ed., Goldendale, WA: Maryhill Museum of Art, 1992. Citations are to the reprint edition.
Turner, James Morton. *The Promise of Wilderness: American Environmental Politics since 1964.* Seattle: University of Washington Press, 2012.
Twight, Ben. *Organizational Values and Political Power: The Forest Service Versus the Olympic National Park.* University Park: Pennsylvania State University Press, 1983.
Twitchell, Heath. *Northwest Epic: The Building of the Alaska Highway.* New York: St. Martin's, 1992.
Twombly, Robert C. *Frank Lloyd Wright: An Interpretive Biography.* New York: Harper & Row, 1973.
Upham, Stedman, and Nat Zappia. *The Many Faces of Edward Sherriff Curtis: Portraits and Stories from Native North America.* Tulsa: Gilcrease Museum and Thomas Gilcrease Association, Seattle: University of Washington Press, 2006.
Watkins, T. H. *Righteous Pilgrim: The Life and Times of Harold L. Ickes, 1874-1952.* New York: Henry Holt, 1990.
White, John H. *The American Railroad Passenger Car.* Baltimore: Johns Hopkins University Press, 1978.
Wik, Reynold M. *Henry Ford and Grass-roots America.* Ann Arbor: University of Michigan Press, 1972.
Wilson, William H. *Railroad in the Clouds: The Alaska Railroad in the Age of Steam, 1914-1945.* Boulder, CO: Pruett, 1977.
Winslow, Kathryn. *Big Pan-Out.* New York: Norton, 1951.
Wood, Robert L. *The Land that Slept Late: The Olympic Mountains in Legend and History.* Seattle: Mountaineers, 1995.

Worster, Donald. *Rivers of Empire: Water, Aridity, and the Growth of the American West.* New York: Oxford University Press, 1985.

Articles

Andrews, Mildred. "Ryther, Mother Ollie (1849-1934)." *HistoryLink.org.* Essay 546.
Bade, William Frederick. "On the Trail with the Sierra Club." *Sierra Club Bulletin* 5 (January 1, 1904): 50-65.
Bailey, Winona. "The Mt. Adams Outing of 1911." *Mountaineer* 4 (1911): 24.
Banks, Mary. "Mountaineers in the Olympics." *Mountaineer* 2 (November 1908): 75-86.
"Battle of Helena." *The Encyclopedia of Arkansas History and Culture*, www.encyclopediaofarkansas.net.
Baule, John. "Last Bite [Ben Perham]." *Good Fruit Grower* (January 1, 2010), online archive.
Betts, William James. "Klondike Photographer." *Alaska Sportsman* 30 (December 1964): 16-19.
Borah, Leo A. "Washington, the Evergreen State: The Amazing Commonwealth of the Pacific Northwest Has Emerged from the Wilderness in a Span of Fifty Years." *National Geographic Magazine* 58 (February 1933): 131-46.
Brooks, William A. "With Sierrans and Mazamas—July 1905." *Appalachia* 11 (May 1906): 114-25.
Burnham, John C. "The Gasoline Tax and the Automobile Revolution." *Mississippi Valley Historical Review* 48 (December 1961): 435-59.
"Climbing Rainier with Curtis in 1909." *Columbia: The Magazine of Northwest History* 1 (Summer 1987): 29-37.
Coulter, Calvin Brewster. "The New Settlers on the Yakima Project." *Pacific Northwest Quarterly* 61 (January 1970): 10-21.
_______. "The Victory of National Irrigation in the Yakima Valley, 1902-1906." *Pacific Northwest Quarterly* 42 (April 1951): 99-122.
Curtis, Asahel. "The First Ascent of Mount Olympus: Climbing the Snow-Clad Olympic Mountains which Overlook the Pacific Ocean." *Worlds' Work* 16 (May 1908): 10, 261-62.
_______. "The First Ascent of Mount Shuksan." *Mountaineer* 1 (June 1907): 52.
_______. "The Future of the Rainier National Park." *Mountaineer* 4 (1911): 42-43.
_______. "The Mountaineers' Annual Outing, 1907." *Mountaineer* 1 (March 1907): 45-47.
_______. "Mountaineers' Outing to Mount Rainier." *Mountaineer* 2 (November 1909): 4-12.
_______. "The Proposed Mount Olympus National Park." *American Forests* 42 (April 1936): 166-69, 195-96.
_______. "Storm Bound on Mount Olympus." *Mountaineer* 1 (September 1907): 64-72.
Curtis, Edward S. "The Rush to the Klondike over the Mountain Passes." *The Century* 55 (March 1898): 692-97.
Danner, Lauren. "Ice Peaks National Park: The Spectacular Failure of a New Deal Idea." *Columbia: The Magazine of Northwest History* 23 (Fall 2009): 29-35.
Duncan, Don. "Curtis, Asahel (1874-1941) Photographer." *HistoryLink.org.* Essay 8780.
Edwards, G. Thomas. "'The Early Morning of Yakima's Day of Greatness': The Yakima County Agricultural Boom of 1905-1911." *Pacific Northwest Quarterly* 73 (April 1982): 78-89.

_______. "Irrigation in Eastern Washington, 1908-1911: The Promotional Photographs of Asahel Curtis." *Pacific Northwest Quarterly* 75 (July 1981): 112-70.

Frederick, Richard. "Asahel Curtis and the Klondike Stampede." *Alaska Journal* 13 (Spring 1983): 113-21.

Hartman, John P. "An Appreciation of the Late Asahel Curtis." *Washington Motorist* 22 (April 1941): 10.

Hong, Suk Chul. "The Burden of Early Exposure to Malaria in the United States, 1850-1860: Malnutrition and Immune Disorders." NIH Public Access, www.ncbi.nlm.nih.gov/pmc/articles/PMC2600412.

King, Molly Leckenby. "Early Outings Through the Eyes of a Girl." *Mountaineer* 50 (December 1956): 36-37.

Kinnick, Chery. "Lawrence Denny Lindsley." *Columbia: The Magazine of Northwest History* 27 (Summer 2013), 3-9.

Lile, Stephanie. "Two Views, Two Voices: The Stereoscopic Perspective of Photographers Asahel and Edward Curtis." *Columbia: The Magazine of Northwest History* 10 (Spring 1996): 17-28.

Nelson, L. A. "The Ascent of Mount Olympus." *Mountaineer* 1 (September 1907): 65-68.

_______. "Thirty Years in Retrospect: The First Decade in Mountaineer Annals." *Mountaineer* 30 (December 1937): 9-12.

Nettleton, Lulie. "Mountaineers' Outing on Glacier Peak." *Mountaineer* 3 (November 1910): 29-40.

Nielsen, Mark. "The Brown Farm on the Nisqually Delta, 1904-1919: a Photographic Essay." *Pacific Northwest Quarterly* 71 (October 1980): 167-71.

MacDonald, Sally. "Image Conscious." *Pacific* [Magazine of the *Seattle Times* and the *Seattle Post-Intelligencer*], June 12, 1988: 16-18, 34-36.

Meany, Edmond S. "The Olympic National Monument." *Mountaineer* 4 (1911): 54-59.

Olson, Alexander I. "Heritage Schemes: The Curtis Brothers and the Indian Moment of Northwest Boosterism." *Western Historical Quarterly* 40 (Summer 2009): 159-78.

Prater, Yvonne. "The Old Yellowstone Trail." *Columbia: The Magazine of Northwest History* 10 (Spring 1996): 39-44.

Ratcliff, Evelyn Marie. "The Sierra Club's Ascent of Mt. Rainier." *Sierra Club Bulletin* 6 (January 1906): 1-5.

"Report of the Outing Committee." *Sierra Club Bulletin* 6 (January 1906): 50-51.

Richardson, Elmo R. "Olympic National Park: Twenty Years of Controversy." *Forest History* 12 (April 1968): 6-15.

Schwantes, Carlos A. "The Milwaukee Road's Pacific Extension, 1909-1929: The Photographs of Asahel Curtis." *Pacific Northwest Quarterly* 71 (January 1981): 30-40.

Weingroff, Richard F. "Federal Aid Road Act of 1916: Building the Foundation." *Public Roads* 60 (Summer 1996).

Wilson, William H. "Asahel Curtis and the Fight over the Olympic National Park." *Pacific Northwest Quarterly* 99 (Summer 2008): 107-21.

_______. "'Names Joined Together as Our Hearts Are': The Friendship of Samuel Hill and Reginald H. Thomson." *Pacific Northwest Quarterly* 94 (Fall 2003): 183-96.

_______. "The War League: Asahel Curtis's Plan for Peace." *Columbia: The Magazine of Northwest History* 23 (Winter 2009-10): 33-35.

Yeon, John B. "The Issue of the Olympics." *American Forests* 42 (June 1936): 255-57, 291-92.

Hearings, Pamphlets, and Reports

Allen, G. F. *Report of the Acting Superintendent of the Mount Rainier National Park to the Secretary of the Interior, 1908.* Washington, DC: 1908.

Lancaster, Samuel C. *Report of Samuel C. Lancaster, Commissioner, Seattle-Tacoma National Park Committee, Services Performed, Sixty-Second Congress, Washington, D.C.* [Seattle: 1913].

A Statement by the "Majority" of the House of Representatives, Extraordinary Session, Legislature, 1925, State of Washington. [Olympia?]: 1925.

U.S. Bureau of the Census. *Seventeenth Decennial Census of the United States, 1950.* Washington, DC: 1950.

U.S. Department of State. *Report of the Commission to Study the Proposed Highway to Alaska 1933.* Washington, DC: 1933.

U.S. Congress. House. Committee on Public Lands. *Mount Olympus National Park: Hearings Before the Committee on the Public Lands.* 74th Cong., 2nd sess., 1936. Washington, DC: 1936.

_______. Committee on Public Lands. *To Establish the Olympic National Park in the State of Washington: Hearings before the Committee on Public Lands.* 75th Cong., 3d sess., 1938. Washington, DC: 1938.

Washington Irrigation Institute, *Proceedings.*

Washington State Department of Transportation Library. *Forty Years with the Washington Department of Highways*, www.wsdot.wa.gov.

Washington State Good Roads Association. *Facts and Figures on Highway Financing and Administration in the State of Washington.* [Seattle]: The Association, 1931.

Washington State Planning Council. *Reclamation, A Sound National Policy: An Inquiry into the Effects of Irrigation Development on Local, State and National Economy As Demonstrated by The Yakima Valley and Other Irrigation Areas in Washington.* [Olympia]: The Council, 1936.

_______. *The Proposed Mount Olympus National Park: A Land Use Study of Public Lands On the Olympic Peninsula.* [Olympia]: The Council, December, 1936.

Washington State Good Roads Association. *Proceedings.*

Waugh, Kathleen, comp. *Roll on Columbia: Guide to the Records of the Columbia Basin Survey and the Columbia Basin Commission.* Olympia: Washington State Archives, 1987.

Newspapers

Aberdeen (WA) *Daily World*
Chehalis (WA) *Bee-Nugget*
New York Herald-Tribune
Seattle Argus
Seattle Business
Seattle Times
Seattle Town Crier
Seattle Post-Intelligencer

Seattle's Business
Spokane Spokesman-Review
Tacoma Ledger
Tacoma News Tribune
Yakima Daily Republic

Theses and Dissertations

Gunns, Albert Francis. "Roland Hill Hartley and the Politics of Washington State." Master's thesis, University of Washington, 1963.
McCoy, Genevieve. "'Call It Mount Tacoma,' A History of the Controversy Over the Name of Mount Rainier." Master's thesis, University of Washington, 1984.
Utter, Kathryn L. "In the End the Land: Settlement of the Columbia Basin Project." Ph.D. diss., University of Washington, 2004.
Wilson, Robert Wayne. "A History of the Washington State Apple Advertising Commission." Master's thesis, University of Washington, 1966.

Court Decisions

State ex. rel. Clausen v. Hartley, 257 Pac. 396.
State ex. rel. Draham v. Cliff Yelle 175 Wash. 33.

Interviews

Curtis, Asahel, Jr., interview by Alex Olson, May 1981, transcribed May 2, 2006. Special Collections, Research Center, Washington State Historical Society, Tacoma.
Lindsley, Lawrence Denny, interview by Harry Majors, 1973. Special Collections, University of Washington Libraries, Seattle.
McCullough, Betty, and Polly Kella, interview by anonymous, August 1982. Special Collections, Research Center, Washington State Historical Society, Tacoma.

Online Sources*

Encyclopedia of Arkansas History and Culture
Good Fruit Grower
HistoryLink
Mazamas
National Civil War Chaplains Museum
National Institutes of Health Public Access
Public Roads
Sierra Club
State Historical Society of Wisconsin
Washington State Department of Transportation
Westegg (inflation data)
Woman's Century Club
Yukon Archives Genealogy Research Database

*These sources are listed for convenience and, in some cases, as online locations for general information about organizations such as the Mazamas, which may be found by using a search engine. Significant institutional sources accessed online are also listed above. Fuller information is available in the relevant chapter notes.

Index

Note: Page numbers in *italic type* indicate illustrations and captions.

About the Author

William H. Wilson is a professor emeritus of history at the University of North Texas and a seasoned researcher and author. He has served as president of the Society for American City and Regional Planning History, and is a member of several other history associations. He received the Lewis Mumford Prize for the Best Book on American Planning History, and the Association of American Publishers' Architecture and Urban Planning Outstanding Book of the Year. His previous titles include *Shaper of Seattle; The City Beautiful Movement;* and *Railroad in the Clouds.*